Activities Workbook
for
Active Calculus Single Variable
Chapters 1–4

Matthew Boelkins
Grand Valley State University

Contributing Authors

David Austin
Grand Valley State University

Steven Schlicker
Grand Valley State University

Production Editor

Mitchel T. Keller
Morningside College

July 26, 2019

Cover Photo: James Haefner Photography

Edition: 2018 Updated

Website: http://activecalculus.org

Preface

This Activities Workbook for *Active Calculus Single Variable* collects all the Preview Activities and Activities in a way that each starts on a new page. The Activities Workbook is designed to be used by students who wish to have a complete set of the activities to work through as they read the book in an electronic format and for instructors who wish to have a one activity per page format to make printing for distribution in class easier.

The design of this workbook is such that each Preview Activity and Activity starts on a right-hand page. As a result, most left-hand pages in this workbook are intentionally left blank as a place for student work associated with one of the adjacent activities. The workbook is offered for purchase in print form in two volumes: the first for chapters 1–4 and the second for chapters 5–8.

Contents

Contents

CHAPTER 1

Understanding the Derivative

1.1 How do we measure velocity?

Preview Activity 1.1.1. Suppose that the height s of a ball at time t (in seconds) is given in feet by the formula $s(t) = 64 - 16(t - 1)^2$.

a. Construct a graph of $y = s(t)$ on the time interval $0 \leq t \leq 3$. Label at least six distinct points on the graph, including the three points showing when the ball was released, when the ball reaches its highest point, and when the ball lands.

b. Describe the behavior of the ball on the time interval $0 < t < 1$ and on time interval $1 < t < 3$. What occurs at the instant $t = 1$?

c. Consider the expression

$$AV_{[0.5,1]} = \frac{s(1) - s(0.5)}{1 - 0.5}.$$

Compute the value of $AV_{[0.5,1]}$. What does this value measure on the graph? What does this value tell us about the motion of the ball? In particular, what are the units on $AV_{[0.5,1]}$?

Activity 1.1.2. The following questions concern the position function given by $s(t) = 64 - 16(t - 1)^2$, considered in Preview Activity 1.1.1.

    a. Compute the average velocity of the ball on each of the following time intervals: $[0.4, 0.8]$, $[0.7, 0.8]$, $[0.79, 0.8]$, $[0.799, 0.8]$, $[0.8, 1.2]$, $[0.8, 0.9]$, $[0.8, 0.81]$, $[0.8, 0.801]$. Include units for each value.

    b. On the graph provided in Figure 1.1.1, sketch the line that passes through the points $A = (0.4, s(0.4))$ and $B = (0.8, s(0.8))$. What is the meaning of the slope of this line? In light of this meaning, what is a geometric way to interpret each of the values computed in the preceding question?

    c. Use a graphing utility to plot the graph of $s(t) = 64 - 16(t - 1)^2$ on an interval containing the value $t = 0.8$. Then, zoom in repeatedly on the point $(0.8, s(0.8))$. What do you observe about how the graph appears as you view it more and more closely?

    d. What do you conjecture is the velocity of the ball at the instant $t = 0.8$? Why?

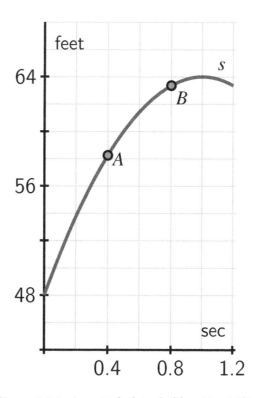

Figure 1.1.1: A partial plot of $s(t) = 64 - 16(t - 1)^2$.

Activity 1.1.3. Each of the following questions concern $s(t) = 64 - 16(t - 1)^2$, the position function from Preview Activity 1.1.1.

    a. Compute the average velocity of the ball on the time interval $[1.5, 2]$. What is different between this value and the average velocity on the interval $[0, 0.5]$?

    b. Use appropriate computing technology to estimate the instantaneous velocity of the ball at $t = 1.5$. Likewise, estimate the instantaneous velocity of the ball at $t = 2$. Which value is greater?

    c. How is the sign of the instantaneous velocity of the ball related to its behavior at a given point in time? That is, what does positive instantaneous velocity tell you the ball is doing? Negative instantaneous velocity?

    d. Without doing any computations, what do you expect to be the instantaneous velocity of the ball at $t = 1$? Why?

Activity 1.1.4. For the function given by $s(t) = 64 - 16(t-1)^2$ from Preview Activity 1.1.1, find the most simplified expression you can for the average velocity of the ball on the interval $[2, 2+h]$. Use your result to compute the average velocity on $[1.5, 2]$ and to estimate the instantaneous velocity at $t = 2$. Finally, compare your earlier work in Activity 1.1.2.

1.2 The notion of limit

Preview Activity 1.2.1. Suppose that g is the function given by the graph below. Use the graph in Figure 1.2.1 to answer each of the following questions.

   a. Determine the values $g(-2)$, $g(-1)$, $g(0)$, $g(1)$, and $g(2)$, if defined. If the function value is not defined, explain what feature of the graph tells you this.

   b. For each of the values $a = -1$, $a = 0$, and $a = 2$, complete the following sentence: "As x gets closer and closer (but not equal) to a, $g(x)$ gets as close as we want to _____."

   c. What happens as x gets closer and closer (but not equal) to $a = 1$? Does the function $g(x)$ get as close as we would like to a single value?

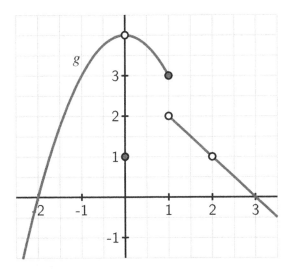

Figure 1.2.1: Graph of $y = g(x)$ for Preview Activity 1.2.1.

Activity 1.2.2. Estimate the value of each of the following limits by constructing appropriate tables of values. Then determine the exact value of the limit by using algebra to simplify the function. Finally, plot each function on an appropriate interval to check your result visually.

a. $\lim_{x \to 1} \frac{x^2-1}{x-1}$

b. $\lim_{x \to 0} \frac{(2+x)^3-8}{x}$

c. $\lim_{x \to 0} \frac{\sqrt{x+1}-1}{x}$

Activity 1.2.3. Consider a moving object whose position function is given by $s(t) = t^2$, where s is measured in meters and t is measured in minutes.

    a. Determine the most simplified expression for the average velocity of the object on the interval $[3, 3 + h]$, where $h > 0$.

    b. Determine the average velocity of the object on the interval $[3, 3.2]$. Include units on your answer.

    c. Determine the instantaneous velocity of the object when $t = 3$. Include units on your answer.

Activity 1.2.4. For the moving object whose position s at time t is given by the graph in Figure 1.2.11, answer each of the following questions. Assume that s is measured in feet and t is measured in seconds.

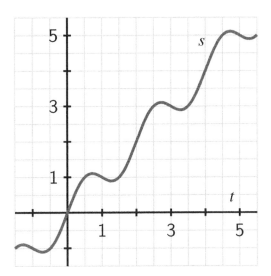

Figure 1.2.11: Plot of the position function $y = s(t)$ in Activity 1.2.4.

a. Use the graph to estimate the average velocity of the object on each of the following intervals: $[0.5, 1]$, $[1.5, 2.5]$, $[0, 5]$. Draw each line whose slope represents the average velocity you seek.

b. How could you use average velocities or slopes of lines to estimate the instantaneous velocity of the object at a fixed time?

c. Use the graph to estimate the instantaneous velocity of the object when $t = 2$. Should this instantaneous velocity at $t = 2$ be greater or less than the average velocity on $[1.5, 2.5]$ that you computed in (a)? Why?

1.3 The derivative of a function at a point

Preview Activity 1.3.1. Suppose that f is the function given by the graph below and that a and $a + h$ are the input values as labeled on the x-axis. Use the graph in Figure 1.3.2 to answer the following questions.

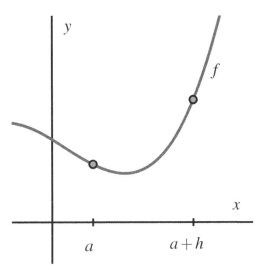

Figure 1.3.2: Plot of $y = f(x)$ for Preview Activity 1.3.1.

a. Locate and label the points $(a, f(a))$ and $(a + h, f(a + h))$ on the graph.

b. Construct a right triangle whose hypotenuse is the line segment from $(a, f(a))$ to $(a + h, f(a + h))$. What are the lengths of the respective legs of this triangle?

c. What is the slope of the line that connects the points $(a, f(a))$ and $(a + h, f(a + h))$?

d. Write a meaningful sentence that explains how the average rate of change of the function on a given interval and the slope of a related line are connected.

Activity 1.3.2. Consider the function f whose formula is $f(x) = 3 - 2x$.

a. What familiar type of function is f? What can you say about the slope of f at every value of x?

b. Compute the average rate of change of f on the intervals $[1, 4]$, $[3, 7]$, and $[5, 5 + h]$; simplify each result as much as possible. What do you notice about these quantities?

c. Use the limit definition of the derivative to compute the exact instantaneous rate of change of f with respect to x at the value $a = 1$. That is, compute $f'(1)$ using the limit definition. Show your work. Is your result surprising?

d. Without doing any additional computations, what are the values of $f'(2)$, $f'(\pi)$, and $f'(-\sqrt{2})$? Why?

Activity 1.3.3. A water balloon is tossed vertically in the air from a window. The balloon's height in feet at time t in seconds after being launched is given by $s(t) = -16t^2 + 16t + 32$. Use this function to respond to each of the following questions.

   a. Sketch an accurate, labeled graph of s on the axes provided in Figure 1.3.10. You should be able to do this without using computing technology.

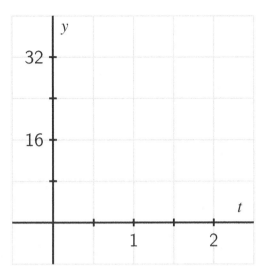

Figure 1.3.10: Axes for plotting $y = s(t)$ in Activity 1.3.3.

   b. Compute the average rate of change of s on the time interval $[1, 2]$. Include units on your answer and write one sentence to explain the meaning of the value you found.

   c. Use the limit definition to compute the instantaneous rate of change of s with respect to time, t, at the instant $a = 1$. Show your work using proper notation, include units on your answer, and write one sentence to explain the meaning of the value you found.

   d. On your graph in (a), sketch two lines: one whose slope represents the average rate of change of s on $[1, 2]$, the other whose slope represents the instantaneous rate of change of s at the instant $a = 1$. Label each line clearly.

   e. For what values of a do you expect $s'(a)$ to be positive? Why? Answer the same questions when "positive" is replaced by "negative" and "zero."

Activity 1.3.4. A rapidly growing city in Arizona has its population P at time t, where t is the number of decades after the year 2010, modeled by the formula $P(t) = 25000e^{t/5}$. Use this function to respond to the following questions.

  a. Sketch an accurate graph of P for $t = 0$ to $t = 5$ on the axes provided in Figure 1.3.11. Label the scale on the axes carefully.

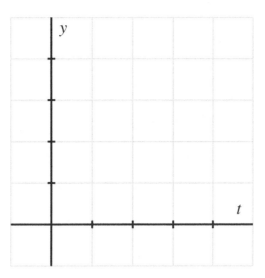

Figure 1.3.11: Axes for plotting $y = P(t)$ in Activity 1.3.4.

  b. Compute the average rate of change of P between 2030 and 2050. Include units on your answer and write one sentence to explain the meaning (in everyday language) of the value you found.

  c. Use the limit definition to write an expression for the instantaneous rate of change of P with respect to time, t, at the instant $a = 2$. Explain why this limit is difficult to evaluate exactly.

  d. Estimate the limit in (c) for the instantaneous rate of change of P at the instant $a = 2$ by using several small h values. Once you have determined an accurate estimate of $P'(2)$, include units on your answer, and write one sentence (using everyday language) to explain the meaning of the value you found.

  e. On your graph above, sketch two lines: one whose slope represents the average rate of change of P on $[2, 4]$, the other whose slope represents the instantaneous rate of change of P at the instant $a = 2$.

  f. In a carefully-worded sentence, describe the behavior of $P'(a)$ as a increases in value. What does this reflect about the behavior of the given function P?

1.4 The derivative function

Preview Activity 1.4.1. Consider the function $f(x) = 4x - x^2$.

a. Use the limit definition to compute the derivative values: $f'(0)$, $f'(1)$, $f'(2)$, and $f'(3)$.

b. Observe that the work to find $f'(a)$ is the same, regardless of the value of a. Based on your work in (a), what do you conjecture is the value of $f'(4)$? How about $f'(5)$? (Note: you should *not* use the limit definition of the derivative to find either value.)

c. Conjecture a formula for $f'(a)$ that depends only on the value a. That is, in the same way that we have a formula for $f(x)$ (recall $f(x) = 4x - x^2$), see if you can use your work above to guess a formula for $f'(a)$ in terms of a.

Activity 1.4.2. For each given graph of $y = f(x)$, sketch an approximate graph of its derivative function, $y = f'(x)$, on the axes immediately below. The scale of the grid for the graph of f is 1×1; assume the horizontal scale of the grid for the graph of f' is identical to that for f. If necessary, adjust and label the vertical scale on the axes for f'.

When you are finished with all 8 graphs, write several sentences that describe your overall process for sketching the graph of the derivative function, given the graph the original function. What are the values of the derivative function that you tend to identify first? What do you do thereafter? How do key traits of the graph of the derivative function exemplify properties of the graph of the original function?

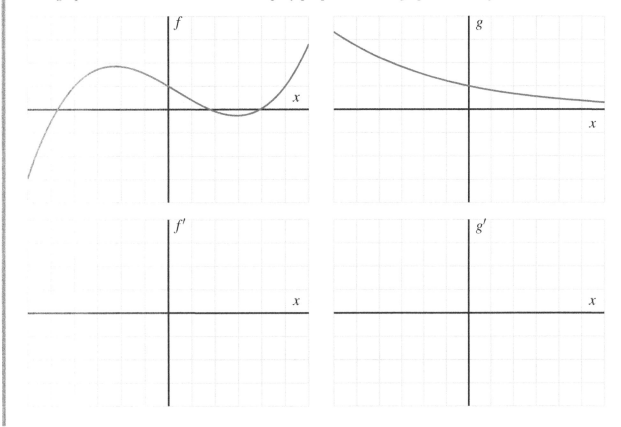

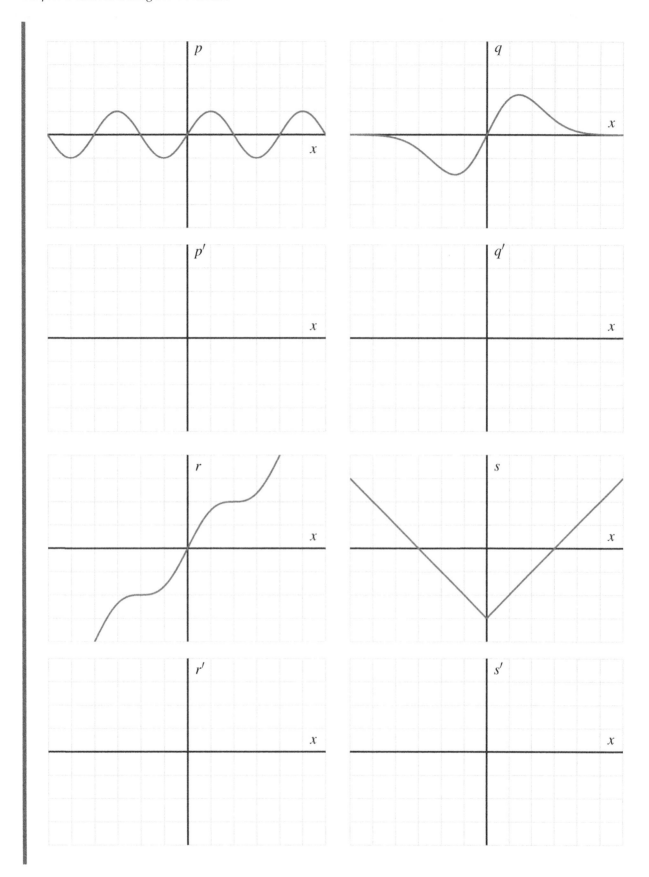

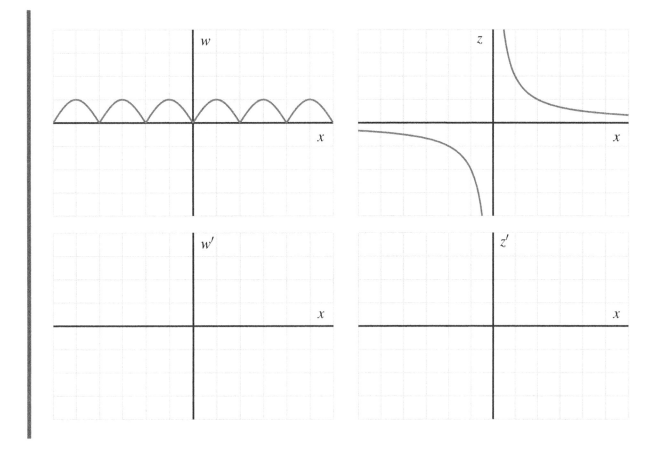

Activity 1.4.3. For each of the listed functions, determine a formula for the derivative function. For the first two, determine the formula for the derivative by thinking about the nature of the given function and its slope at various points; do not use the limit definition. For the latter four, use the limit definition. Pay careful attention to the function names and independent variables. It is important to be comfortable with using letters other than f and x. For example, given a function $p(z)$, we call its derivative $p'(z)$.

a. $f(x) = 1$

b. $g(t) = t$

c. $p(z) = z^2$

d. $q(s) = s^3$

e. $F(t) = \frac{1}{t}$

f. $G(y) = \sqrt{y}$

1.5 Interpreting, estimating, and using the derivative

Preview Activity 1.5.1. One of the longest stretches of straight (and flat) road in North America can be found on the Great Plains in the state of North Dakota on state highway 46, which lies just south of the interstate highway I-94 and runs through the town of Gackle. A car leaves town (at time $t = 0$) and heads east on highway 46; its position in miles from Gackle at time t in minutes is given by the graph of the function in Figure 1.5.1. Three important points are labeled on the graph; where the curve looks linear, assume that it is indeed a straight line.

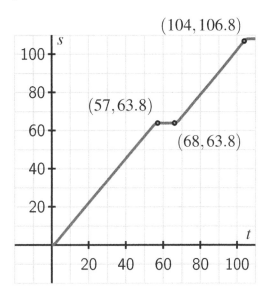

Figure 1.5.1: The graph of $y = s(t)$, the position of the car along highway 46, which tells its distance in miles from Gackle, ND, at time t in minutes.

a. In everyday language, describe the behavior of the car over the provided time interval. In particular, discuss what is happening on the time intervals $[57, 68]$ and $[68, 104]$.

b. Find the slope of the line between the points $(57, 63.8)$ and $(104, 106.8)$. What are the units on this slope? What does the slope represent?

c. Find the average rate of change of the car's position on the interval $[68, 104]$. Include units on your answer.

d. Estimate the instantaneous rate of change of the car's position at the moment $t = 80$. Write a sentence to explain your reasoning and the meaning of this value.

1.5 Interpreting, estimating, and using the derivative

Activity 1.5.2. A potato is placed in an oven, and the potato's temperature F (in degrees Fahrenheit) at various points in time is taken and recorded in the following table. Time t is measured in minutes.

t	0	15	30	45	60	75	90
$F(t)$	70	180.5	251	296	324.5	342.8	354.5

Table 1.5.4: Temperature data in degrees Fahrenheit.

a. Use a central difference to estimate the instantaneous rate of change of the temperature of the potato at $t = 30$. Include units on your answer.

b. Use a central difference to estimate the instantaneous rate of change of the temperature of the potato at $t = 60$. Include units on your answer.

c. Without doing any calculation, which do you expect to be greater: $F'(75)$ or $F'(90)$? Why?

d. Suppose it is given that $F(64) = 330.28$ and $F'(64) = 1.341$. What are the units on these two quantities? What do you expect the temperature of the potato to be when $t = 65$? when $t = 66$? Why?

e. Write a couple of careful sentences that describe the behavior of the temperature of the potato on the time interval $[0, 90]$, as well as the behavior of the instantaneous rate of change of the temperature of the potato on the same time interval.

Activity 1.5.3. A company manufactures rope, and the total cost of producing r feet of rope is $C(r)$ dollars.

a. What does it mean to say that $C(2000) = 800$?

b. What are the units of $C'(r)$?

c. Suppose that $C(2000) = 800$ and $C'(2000) = 0.35$. Estimate $C(2100)$, and justify your estimate by writing at least one sentence that explains your thinking.

d. Do you think $C'(2000)$ is less than, equal to, or greater than $C'(3000)$? Why?

e. Suppose someone claims that $C'(5000) = -0.1$. What would the practical meaning of this derivative value tell you about the approximate cost of the next foot of rope? Is this possible? Why or why not?

Activity 1.5.4. Researchers at a major car company have found a function that relates gasoline consumption to speed for a particular model of car. In particular, they have determined that the consumption C, in *liters per kilometer*, at a given speed s, is given by a function $C = f(s)$, where s is the car's speed in *kilometers per hour*.

    a. Data provided by the car company tells us that $f(80) = 0.015$, $f(90) = 0.02$, and $f(100) = 0.027$. Use this information to estimate the instantaneous rate of change of fuel consumption with respect to speed at $s = 90$. Be as accurate as possible, use proper notation, and include units on your answer.

    b. By writing a complete sentence, interpret the meaning (in the context of fuel consumption) of "$f(80) = 0.015$."

    c. Write at least one complete sentence that interprets the meaning of the value of $f'(90)$ that you estimated in (a).

1.6 The second derivative

Preview Activity 1.6.1. The position of a car driving along a straight road at time t in minutes is given by the function $y = s(t)$ that is pictured in Figure 1.6.2. The car's position function has units measured in thousands of feet. For instance, the point $(2, 4)$ on the graph indicates that after 2 minutes, the car has traveled 4000 feet.

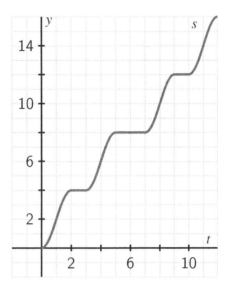

Figure 1.6.2: The graph of $y = s(t)$, the position of the car (measured in thousands of feet from its starting location) at time t in minutes.

a. In everyday language, describe the behavior of the car over the provided time interval. In particular, you should carefully discuss what is happening on each of the time intervals $[0, 1], [1, 2], [2, 3], [3, 4]$, and $[4, 5]$, plus provide commentary overall on what the car is doing on the interval $[0, 12]$.

b. On the lefthand axes provided in Figure 1.6.3, sketch a careful, accurate graph of $y = s'(t)$.

c. What is the meaning of the function $y = s'(t)$ in the context of the given problem? What can we say about the car's behavior when $s'(t)$ is positive? when $s'(t)$ is zero? when $s'(t)$ is negative?

d. Rename the function you graphed in (b) to be called $y = v(t)$. Describe the behavior of v in words, using phrases like "v is increasing on the interval ..." and "v is constant on the interval ...."

e. Sketch a graph of the function $y = v'(t)$ on the righthand axes provide in Figure 1.6.3. Write at least one sentence to explain how the behavior of $v'(t)$ is connected to the graph of $y = v(t)$.

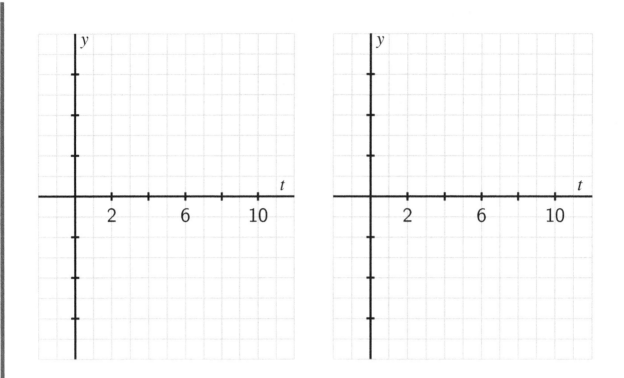

Figure 1.6.3: Axes for plotting $y = v(t) = s'(t)$ and $y = v'(t)$.

Activity 1.6.2. The position of a car driving along a straight road at time t in minutes is given by the function $y = s(t)$ that is pictured in Figure 1.6.11. The car's position function has units measured in thousands of feet. Remember that you worked with this function and sketched graphs of $y = v(t) = s'(t)$ and $y = v'(t)$ in Preview Activity 1.6.1.

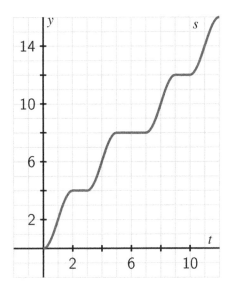

Figure 1.6.11: The graph of $y = s(t)$, the position of the car (measured in thousands of feet from its starting location) at time t in minutes.

a. On what intervals is the position function $y = s(t)$ increasing? decreasing? Why?

b. On which intervals is the velocity function $y = v(t) = s'(t)$ increasing? decreasing? neither? Why?

c. *Acceleration* is defined to be the instantaneous rate of change of velocity, as the acceleration of an object measures the rate at which the velocity of the object is changing. Say that the car's acceleration function is named $a(t)$. How is $a(t)$ computed from $v(t)$? How is $a(t)$ computed from $s(t)$? Explain.

d. What can you say about s'' whenever s' is increasing? Why?

e. Using only the words *increasing, decreasing, constant, concave up, concave down,* and *linear,* complete the following sentences. For the position function s with velocity v and acceleration a,

   - on an interval where v is positive, s is _____.
   - on an interval where v is negative, s is _____.
   - on an interval where v is zero, s is _____.
   - on an interval where a is positive, v is _____.
   - on an interval where a is negative, v is _____.
   - on an interval where a is zero, v is _____.
   - on an interval where a is positive, s is _____.
   - on an interval where a is negative, s is _____.
   - on an interval where a is zero, s is _____.

Activity 1.6.3. A potato is placed in an oven, and the potato's temperature F (in degrees Fahrenheit) at various points in time is taken and recorded in the following table. Time t is measured in minutes. In Activity 1.5.2, we computed approximations to $F'(30)$ and $F'(60)$ using central differences. Those values and more are provided in the second table below, along with several others computed in the same way.

t	$F(t)$
0	70
15	180.5
30	251
45	296
60	324.5
75	342.8
90	354.5

Table 1.6.12: Select values of $F(t)$.

t	$F'(t)$
0	NA
15	6.03
30	3.85
45	2.45
60	1.56
75	1.00
90	NA

Table 1.6.13: Select values of $F'(t)$.

a. What are the units on the values of $F'(t)$?

b. Use a central difference to estimate the value of $F''(30)$.

c. What is the meaning of the value of $F''(30)$ that you have computed in (b) in terms of the potato's temperature? Write several careful sentences that discuss, with appropriate units, the values of $F(30)$, $F'(30)$, and $F''(30)$, and explain the overall behavior of the potato's temperature at this point in time.

d. Overall, is the potato's temperature increasing at an increasing rate, increasing at a constant rate, or increasing at a decreasing rate? Why?

Activity 1.6.4. This activity builds on our experience and understanding of how to sketch the graph of f' given the graph of f.

In Figure 1.6.14, given the respective graphs of two different functions f, sketch the corresponding graph of f' on the first axes below, and then sketch f'' on the second set of axes. In addition, for each, write several careful sentences in the spirit of those in Activity 1.6.2 that connect the behaviors of f, f', and f''. For instance, write something such as

f' is _____ on the interval _____, which is connected to the fact that f is _____ on the same interval _____, and f'' is _____ on the interval.

but of course with the blanks filled in. Throughout, view the scale of the grid for the graph of f as being 1×1, and assume the horizontal scale of the grid for the graph of f' is identical to that for f. If you need to adjust the vertical scale on the axes for the graph of f' or f'', you should label that accordingly.

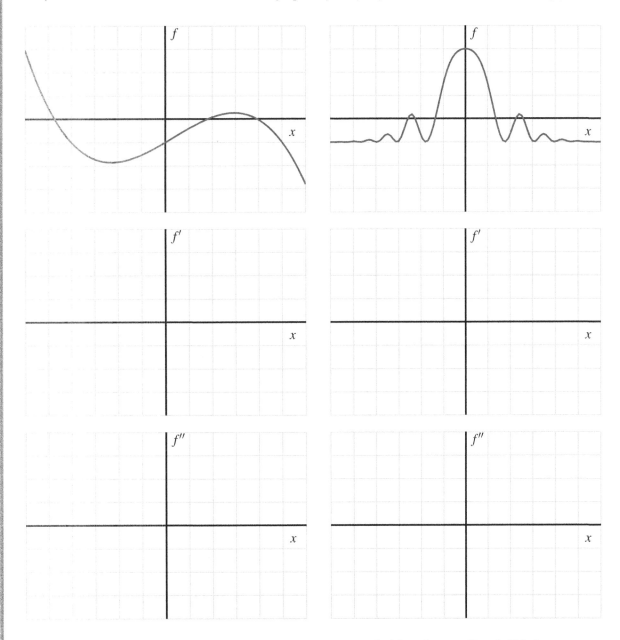

Figure 1.6.14: Two given functions f, with axes provided for plotting f' and f'' below.

1.7 Limits, Continuity, and Differentiability

Preview Activity 1.7.1. A function f defined on $-4 < x < 4$ is given by the graph in Figure 1.7.1. Use the graph to answer each of the following questions. Note: to the right of $x = 2$, the graph of f is exhibiting infinite oscillatory behavior similar to the function $\sin(\frac{\pi}{x})$ that we encountered in the key example early in Section 1.2.

a. For each of the values $a = -3, -2, -1, 0, 1, 2, 3$, determine whether or not $\lim_{x \to a} f(x)$ exists. If the function has a limit L at a given point, state the value of the limit using the notation $\lim_{x \to a} f(x) = L$. If the function does not have a limit at a given point, write a sentence to explain why.

b. For each of the values of a from part (a) where f has a limit, determine the value of $f(a)$ at each such point. In addition, for each such a value, does $f(a)$ have the same value as $\lim_{x \to a} f(x)$?

c. For each of the values $a = -3, -2, -1, 0, 1, 2, 3$, determine whether or not $f'(a)$ exists. In particular, based on the given graph, ask yourself if it is reasonable to say that f has a tangent line at $(a, f(a))$ for each of the given a-values. If so, visually estimate the slope of the tangent line to find the value of $f'(a)$.

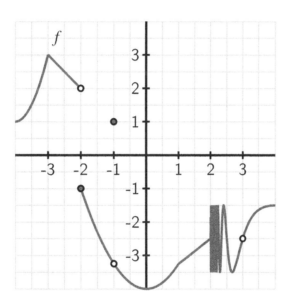

Figure 1.7.1: The graph of $y = f(x)$.

Activity 1.7.2. Consider a function that is piecewise-defined according to the formula

$$f(x) = \begin{cases} 3(x+2)+2 & \text{for } -3 < x < -2 \\ \frac{2}{3}(x+2)+1 & \text{for } -2 \le x < -1 \\ \frac{2}{3}(x+2)+1 & \text{for } -1 < x < 1 \\ 2 & \text{for } x = 1 \\ 4-x & \text{for } x > 1 \end{cases}$$

Use the given formula to answer the following questions.

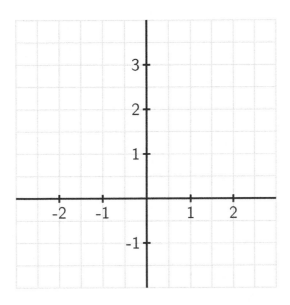

Figure 1.7.4: Axes for plotting the function $y = f(x)$ in Activity 1.7.2.

a. For each of the values $a = -2, -1, 0, 1, 2$, compute $f(a)$.

b. For each of the values $a = -2, -1, 0, 1, 2$, determine $\lim_{x \to a^-} f(x)$ and $\lim_{x \to a^+} f(x)$.

c. For each of the values $a = -2, -1, 0, 1, 2$, determine $\lim_{x \to a} f(x)$. If the limit fails to exist, explain why by discussing the left- and right-hand limits at the relevant a-value.

d. For which values of a is the following statement true?

$$\lim_{x \to a} f(x) \ne f(a)$$

e. On the axes provided in Figure 1.7.4, sketch an accurate, labeled graph of $y = f(x)$. Be sure to carefully use open circles (○) and filled circles (●) to represent key points on the graph, as dictated by the piecewise formula.

Activity 1.7.3. This activity builds on your work in Preview Activity 1.7.1, using the same function f as given by the graph that is repeated in Figure 1.7.7.

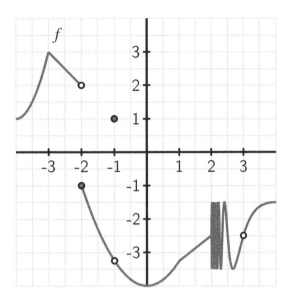

Figure 1.7.7: The graph of $y = f(x)$ for Activity 1.7.3.

a. At which values of a does $\lim_{x \to a} f(x)$ not exist?

b. At which values of a is $f(a)$ not defined?

c. At which values of a does f have a limit, but $\lim_{x \to a} f(x) \neq f(a)$?

d. State all values of a for which f is not continuous at $x = a$.

e. Which condition is stronger, and hence implies the other: f has a limit at $x = a$ or f is continuous at $x = a$? Explain, and hence complete the following sentence: "If f _____ at $x = a$, then f _____ at $x = a$," where you complete the blanks with *has a limit* and *is continuous*, using each phrase once.

Activity 1.7.4. In this activity, we explore two different functions and classify the points at which each is not differentiable. Let g be the function given by the rule $g(x) = |x|$, and let f be the function that we have previously explored in Preview Activity 1.7.1, whose graph is given again in Figure 1.7.9.

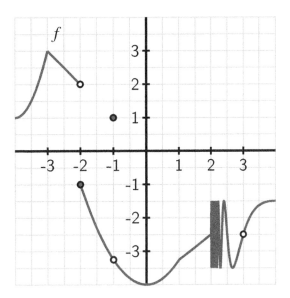

Figure 1.7.9: The graph of $y = f(x)$ for Activity 1.7.4.

a. Reasoning visually, explain why g is differentiable at every point x such that $x \neq 0$.

b. Use the limit definition of the derivative to show that $g'(0) = \lim_{h \to 0} \frac{|h|}{h}$.

c. Explain why $g'(0)$ fails to exist by using small positive and negative values of h.

d. State all values of a for which f is not differentiable at $x = a$. For each, provide a reason for your conclusion.

e. True or false: if a function p is differentiable at $x = b$, then $\lim_{x \to b} p(x)$ must exist. Why?

1.8 The Tangent Line Approximation

Preview Activity 1.8.1. Consider the function $y = g(x) = -x^2 + 3x + 2$.

  a. Use the limit definition of the derivative to compute a formula for $y = g'(x)$.

  b. Determine the slope of the tangent line to $y = g(x)$ at the value $x = 2$.

  c. Compute $g(2)$.

  d. Find an equation for the tangent line to $y = g(x)$ at the point $(2, g(2))$. Write your result in point-slope form.

  e. On the axes provided in Figure 1.8.1, sketch an accurate, labeled graph of $y = g(x)$ along with its tangent line at the point $(2, g(2))$.

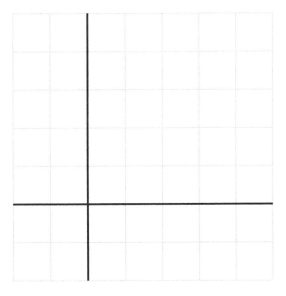

Figure 1.8.1: Axes for plotting $y = g(x)$ and its tangent line to the point $(2, g(2))$.

Activity 1.8.2. Suppose it is known that for a given differentiable function $y = g(x)$, its local linearization at the point where $a = -1$ is given by $L(x) = -2 + 3(x + 1)$.

   a. Compute the values of $L(-1)$ and $L'(-1)$.

   b. What must be the values of $g(-1)$ and $g'(-1)$? Why?

   c. Do you expect the value of $g(-1.03)$ to be greater than or less than the value of $g(-1)$? Why?

   d. Use the local linearization to estimate the value of $g(-1.03)$.

   e. Suppose that you also know that $g''(-1) = 2$. What does this tell you about the graph of $y = g(x)$ at $a = -1$?

   f. For x near -1, sketch the graph of the local linearization $y = L(x)$ as well as a possible graph of $y = g(x)$ on the axes provided in Figure 1.8.4.

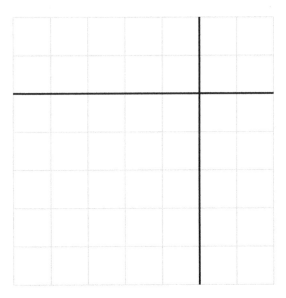

Figure 1.8.4: Axes for plotting $y = L(x)$ and $y = g(x)$.

Activity 1.8.3. This activity concerns a function $f(x)$ about which the following information is known:

- f is a differentiable function defined at every real number x

- $f(2) = -1$

- $y = f'(x)$ has its graph given in Figure 1.8.6

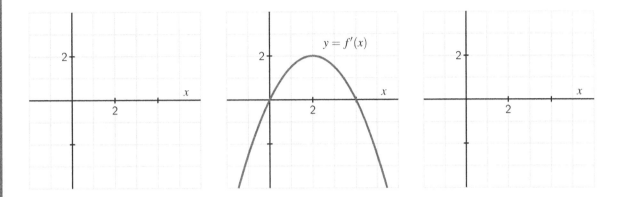

Figure 1.8.6: At center, a graph of $y = f'(x)$; at left, axes for plotting $y = f(x)$; at right, axes for plotting $y = f''(x)$.

Your task is to determine as much information as possible about f (especially near the value $a = 2$) by responding to the questions below.

a. Find a formula for the tangent line approximation, $L(x)$, to f at the point $(2, -1)$.

b. Use the tangent line approximation to estimate the value of $f(2.07)$. Show your work carefully and clearly.

c. Sketch a graph of $y = f''(x)$ on the righthand grid in Figure 1.8.6; label it appropriately.

d. Is the slope of the tangent line to $y = f(x)$ increasing, decreasing, or neither when $x = 2$? Explain.

e. Sketch a possible graph of $y = f(x)$ near $x = 2$ on the lefthand grid in Figure 1.8.6. Include a sketch of $y = L(x)$ (found in part (a)). Explain how you know the graph of $y = f(x)$ looks like you have drawn it.

f. Does your estimate in (b) over- or under-estimate the true value of $f(2.07)$? Why?

CHAPTER 2

Computing Derivatives

2.1 Elementary derivative rules

Preview Activity 2.1.1. Functions of the form $f(x) = x^n$, where $n = 1, 2, 3, \ldots$, are often called **power functions**. The first two questions below revisit work we did earlier in Chapter 1, and the following questions extend those ideas to higher powers of x.

a. Use the limit definition of the derivative to find $f'(x)$ for $f(x) = x^2$.

b. Use the limit definition of the derivative to find $f'(x)$ for $f(x) = x^3$.

c. Use the limit definition of the derivative to find $f'(x)$ for $f(x) = x^4$. (Hint: $(a + b)^4 = a^4 + 4a^3b + 6a^2b^2 + 4ab^3 + b^4$. Apply this rule to $(x + h)^4$ within the limit definition.)

d. Based on your work in (a), (b), and (c), what do you conjecture is the derivative of $f(x) = x^5$? Of $f(x) = x^{13}$?

e. Conjecture a formula for the derivative of $f(x) = x^n$ that holds for any positive integer n. That is, given $f(x) = x^n$ where n is a positive integer, what do you think is the formula for $f'(x)$?

Activity 2.1.2. Use the three rules above to determine the derivative of each of the following functions. For each, state your answer using full and proper notation, labeling the derivative with its name. For example, if you are given a function $h(z)$, you should write "$h'(z) =$" or "$\frac{dh}{dz} =$" as part of your response.

a. $f(t) = \pi$

b. $g(z) = 7^z$

c. $h(w) = w^{3/4}$

d. $p(x) = 3^{1/2}$

e. $r(t) = (\sqrt{2})^t$

f. $s(q) = q^{-1}$

g. $m(t) = \frac{1}{t^3}$

Activity 2.1.3. Use only the rules for constant, power, and exponential functions, together with the Constant Multiple and Sum Rules, to compute the derivative of each function below with respect to the given independent variable. Note well that we do not yet know any rules for how to differentiate the product or quotient of functions. This means that you may have to do some algebra first on the functions below before you can actually use existing rules to compute the desired derivative formula. In each case, label the derivative you calculate with its name using proper notation such as $f'(x)$, $h'(z)$, dr/dt, etc.

a. $f(x) = x^{5/3} - x^4 + 2^x$

b. $g(x) = 14e^x + 3x^5 - x$

c. $h(z) = \sqrt{z} + \frac{1}{z^4} + 5^z$

d. $r(t) = \sqrt{53}\, t^7 - \pi e^t + e^4$

e. $s(y) = (y^2 + 1)(y^2 - 1)$

f. $q(x) = \frac{x^3 - x + 2}{x}$

g. $p(a) = 3a^4 - 2a^3 + 7a^2 - a + 12$

Activity 2.1.4. Each of the following questions asks you to use derivatives to answer key questions about functions. Be sure to think carefully about each question and to use proper notation in your responses.

a. Find the slope of the tangent line to $h(z) = \sqrt{z} + \frac{1}{z}$ at the point where $z = 4$.

b. A population of cells is growing in such a way that its total number in millions is given by the function $P(t) = 2(1.37)^t + 32$, where t is measured in days.

   i. Determine the instantaneous rate at which the population is growing on day 4, and include units on your answer.

   ii. Is the population growing at an increasing rate or growing at a decreasing rate on day 4? Explain.

c. Find an equation for the tangent line to the curve $p(a) = 3a^4 - 2a^3 + 7a^2 - a + 12$ at the point where $a = -1$.

d. What is the difference between being asked to find the *slope* of the tangent line (asked in (a)) and the *equation* of the tangent line (asked in (c))?

2.2 The sine and cosine functions

Preview Activity 2.2.1. Consider the function $g(x) = 2^x$, which is graphed in Figure 2.2.1.

a. At each of $x = -2, -1, 0, 1, 2$, use a straightedge to sketch an accurate tangent line to $y = g(x)$.

b. Use the provided grid to estimate the slope of the tangent line you drew at each point in (a).

c. Use the limit definition of the derivative to estimate $g'(0)$ by using small values of h, and compare the result to your visual estimate for the slope of the tangent line to $y = g(x)$ at $x = 0$ in (b).

d. Based on your work in (a), (b), and (c), sketch an accurate graph of $y = g'(x)$ on the axes adjacent to the graph of $y = g(x)$.

e. Write at least one sentence that explains why it is reasonable to think that $g'(x) = cg(x)$, where c is a constant. In addition, calculate $\ln(2)$, and then discuss how this value, combined with your work above, reasonably suggests that $g'(x) = 2^x \ln(2)$.

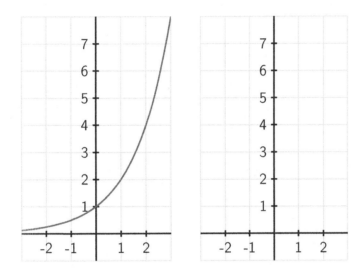

Figure 2.2.1: At left, the graph of $y = g(x) = 2^x$. At right, axes for plotting $y = g'(x)$.

Activity 2.2.2. Consider the function $f(x) = \sin(x)$, which is graphed in Figure 2.2.2 below. Note carefully that the grid in the diagram does not have boxes that are 1×1, but rather approximately 1.57×1, as the horizontal scale of the grid is $\pi/2$ units per box.

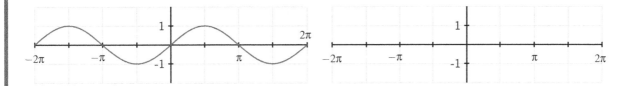

Figure 2.2.2: At left, the graph of $y = f(x) = \sin(x)$.

a. At each of $x = -2\pi, -\frac{3\pi}{2}, -\pi, -\frac{\pi}{2}, 0, \frac{\pi}{2}, \pi, \frac{3\pi}{2}, 2\pi$, use a straightedge to sketch an accurate tangent line to $y = f(x)$.

b. Use the provided grid to estimate the slope of the tangent line you drew at each point. Pay careful attention to the scale of the grid.

c. Use the limit definition of the derivative to estimate $f'(0)$ by using small values of h, and compare the result to your visual estimate for the slope of the tangent line to $y = f(x)$ at $x = 0$ in (b). Using periodicity, what does this result suggest about $f'(2\pi)$? about $f'(-2\pi)$?

d. Based on your work in (a), (b), and (c), sketch an accurate graph of $y = f'(x)$ on the axes adjacent to the graph of $y = f(x)$.

e. What familiar function do you think is the derivative of $f(x) = \sin(x)$?

Activity 2.2.3. Consider the function $g(x) = \cos(x)$, which is graphed in Figure 2.2.5 below. Note carefully that the grid in the diagram does not have boxes that are 1×1, but rather approximately 1.57×1, as the horizontal scale of the grid is $\pi/2$ units per box.

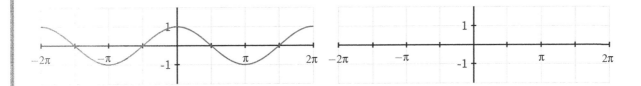

Figure 2.2.5: At left, the graph of $y = g(x) = \cos(x)$.

a. At each of $x = -2\pi, -\frac{3\pi}{2}, -\pi, -\frac{\pi}{2}, 0, \frac{\pi}{2}, \pi, \frac{3\pi}{2}, 2\pi$, use a straightedge to sketch an accurate tangent line to $y = g(x)$.

b. Use the provided grid to estimate the slope of the tangent line you drew at each point. Again, note the scale of the axes and grid.

c. Use the limit definition of the derivative to estimate $g'(\frac{\pi}{2})$ by using small values of h, and compare the result to your visual estimate for the slope of the tangent line to $y = g(x)$ at $x = \frac{\pi}{2}$ in (b). Using periodicity, what does this result suggest about $g'(-\frac{3\pi}{2})$? can symmetry on the graph help you estimate other slopes easily?

d. Based on your work in (a), (b), and (c), sketch an accurate graph of $y = g'(x)$ on the axes adjacent to the graph of $y = g(x)$.

e. What familiar function do you think is the derivative of $g(x) = \cos(x)$?

75

Activity 2.2.4. Answer each of the following questions. Where a derivative is requested, be sure to label the derivative function with its name using proper notation.

a. Determine the derivative of $h(t) = 3\cos(t) - 4\sin(t)$.

b. Find the exact slope of the tangent line to $y = f(x) = 2x + \frac{\sin(x)}{2}$ at the point where $x = \frac{\pi}{6}$.

c. Find the equation of the tangent line to $y = g(x) = x^2 + 2\cos(x)$ at the point where $x = \frac{\pi}{2}$.

d. Determine the derivative of $p(z) = z^4 + 4^z + 4\cos(z) - \sin(\frac{\pi}{2})$.

e. The function $P(t) = 24 + 8\sin(t)$ represents a population of a particular kind of animal that lives on a small island, where P is measured in hundreds and t is measured in decades since January 1, 2010. What is the instantaneous rate of change of P on January 1, 2030? What are the units of this quantity? Write a sentence in everyday language that explains how the population is behaving at this point in time.

2.3 The product and quotient rules

Preview Activity 2.3.1. Let f and g be the functions defined by $f(t) = 2t^2$ and $g(t) = t^3 + 4t$.

a. Determine $f'(t)$ and $g'(t)$.

b. Let $p(t) = 2t^2(t^3 + 4t)$ and observe that $p(t) = f(t) \cdot g(t)$. Rewrite the formula for p by distributing the $2t^2$ term. Then, compute $p'(t)$ using the sum and constant multiple rules.

c. True or false: $p'(t) = f'(t) \cdot g'(t)$.

d. Let $q(t) = \frac{t^3 + 4t}{2t^2}$ and observe that $q(t) = \frac{g(t)}{f(t)}$. Rewrite the formula for q by dividing each term in the numerator by the denominator and simplify to write q as a sum of constant multiples of powers of t. Then, compute $q'(t)$ using the sum and constant multiple rules.

e. True or false: $q'(t) = \frac{g'(t)}{f'(t)}$.

Activity 2.3.2. Use the product rule to answer each of the questions below. Throughout, be sure to carefully label any derivative you find by name. It is not necessary to algebraically simplify any of the derivatives you compute.

a. Let $m(w) = 3w^{17}4^w$. Find $m'(w)$.

b. Let $h(t) = (\sin(t) + \cos(t))t^4$. Find $h'(t)$.

c. Determine the slope of the tangent line to the curve $y = f(x)$ at the point where $a = 1$ if f is given by the rule $f(x) = e^x \sin(x)$.

d. Find the tangent line approximation $L(x)$ to the function $y = g(x)$ at the point where $a = -1$ if g is given by the rule $g(x) = (x^2 + x)2^x$.

Activity 2.3.3. Use the quotient rule to answer each of the questions below. Throughout, be sure to carefully label any derivative you find by name. That is, if you're given a formula for $f(x)$, clearly label the formula you find for $f'(x)$. It is not necessary to algebraically simplify any of the derivatives you compute.

a. Let $r(z) = \frac{3^z}{z^4+1}$. Find $r'(z)$.

b. Let $v(t) = \frac{\sin(t)}{\cos(t)+t^2}$. Find $v'(t)$.

c. Determine the slope of the tangent line to the curve $R(x) = \frac{x^2 - 2x - 8}{x^2 - 9}$ at the point where $x = 0$.

d. When a camera flashes, the intensity I of light seen by the eye is given by the function

$$I(t) = \frac{100t}{e^t},$$

where I is measured in candles and t is measured in milliseconds. Compute $I'(0.5)$, $I'(2)$, and $I'(5)$; include appropriate units on each value; and discuss the meaning of each.

Activity 2.3.4. Use relevant derivative rules to answer each of the questions below. Throughout, be sure to use proper notation and carefully label any derivative you find by name.

a. Let $f(r) = (5r^3 + \sin(r))(4^r - 2\cos(r))$. Find $f'(r)$.

b. Let $p(t) = \dfrac{\cos(t)}{t^6 \cdot 6^t}$. Find $p'(t)$.

c. Let $g(z) = 3z^7 e^z - 2z^2 \sin(z) + \frac{z}{z^2+1}$. Find $g'(z)$.

d. A moving particle has its position in feet at time t in seconds given by the function $s(t) = \frac{3\cos(t)-\sin(t)}{e^t}$. Find the particle's instantaneous velocity at the moment $t = 1$.

e. Suppose that $f(x)$ and $g(x)$ are differentiable functions and it is known that $f(3) = -2$, $f'(3) = 7$, $g(3) = 4$, and $g'(3) = -1$. If $p(x) = f(x) \cdot g(x)$ and $q(x) = \dfrac{f(x)}{g(x)}$, calculate $p'(3)$ and $q'(3)$.

2.4 Derivatives of other trigonometric functions

Preview Activity 2.4.1. Consider the function $f(x) = \tan(x)$, and remember that $\tan(x) = \frac{\sin(x)}{\cos(x)}$.

a. What is the domain of f?

b. Use the quotient rule to show that one expression for $f'(x)$ is

$$f'(x) = \frac{\cos(x)\cos(x) + \sin(x)\sin(x)}{\cos^2(x)}.$$

c. What is the Fundamental Trigonometric Identity? How can this identity be used to find a simpler form for $f'(x)$?

d. Recall that $\sec(x) = \frac{1}{\cos(x)}$. How can we express $f'(x)$ in terms of the secant function?

e. For what values of x is $f'(x)$ defined? How does this set compare to the domain of f?

Activity 2.4.2. Let $h(x) = \sec(x)$ and recall that $\sec(x) = \frac{1}{\cos(x)}$.

    a. What is the domain of h?

    b. Use the quotient rule to develop a formula for $h'(x)$ that is expressed completely in terms of $\sin(x)$ and $\cos(x)$.

    c. How can you use other relationships among trigonometric functions to write $h'(x)$ only in terms of $\tan(x)$ and $\sec(x)$?

    d. What is the domain of h'? How does this compare to the domain of h?

Activity 2.4.3. Let $p(x) = \csc(x)$ and recall that $\csc(x) = \frac{1}{\sin(x)}$.

a. What is the domain of p?

b. Use the quotient rule to develop a formula for $p'(x)$ that is expressed completely in terms of $\sin(x)$ and $\cos(x)$.

c. How can you use other relationships among trigonometric functions to write $p'(x)$ only in terms of $\cot(x)$ and $\csc(x)$?

d. What is the domain of p'? How does this compare to the domain of p?

Activity 2.4.4. Answer each of the following questions. Where a derivative is requested, be sure to label the derivative function with its name using proper notation.

a. Let $f(x) = 5\sec(x) - 2\csc(x)$. Find the slope of the tangent line to f at the point where $x = \frac{\pi}{3}$.

b. Let $p(z) = z^2 \sec(z) - z\cot(z)$. Find the instantaneous rate of change of p at the point where $z = \frac{\pi}{4}$.

c. Let $h(t) = \dfrac{\tan(t)}{t^2 + 1} - 2e^t \cos(t)$. Find $h'(t)$.

d. Let $g(r) = \dfrac{r\sec(r)}{5^r}$. Find $g'(r)$.

e. When a mass hangs from a spring and is set in motion, the object's position oscillates in a way that the size of the oscillations decrease. This is usually called a *damped oscillation*. Suppose that for a particular object, its displacement from equilibrium (where the object sits at rest) is modeled by the function

$$s(t) = \frac{15\sin(t)}{e^t}.$$

Assume that s is measured in inches and t in seconds. Sketch a graph of this function for $t \geq 0$ to see how it represents the situation described. Then compute ds/dt, state the units on this function, and explain what it tells you about the object's motion. Finally, compute and interpret $s'(2)$.

2.5 The chain rule

Preview Activity 2.5.1. For each function given below, identify its fundamental algebraic structure. In particular, is the given function a sum, product, quotient, or composition of basic functions? If the function is a composition of basic functions, state a formula for the inner function g and the outer function f so that the overall composite function can be written in the form $f(g(x))$. If the function is a sum, product, or quotient of basic functions, use the appropriate rule to determine its derivative.

a. $h(x) = \tan(2^x)$

b. $p(x) = 2^x \tan(x)$

c. $r(x) = (\tan(x))^2$

d. $m(x) = e^{\tan(x)}$

e. $w(x) = \sqrt{x} + \tan(x)$

f. $z(x) = \sqrt{\tan(x)}$

Activity 2.5.2. For each function given below, identify an inner function g and outer function f to write the function in the form $f(g(x))$. Determine $f'(x)$, $g'(x)$, and $f'(g(x))$, and then apply the chain rule to determine the derivative of the given function.

a. $h(x) = \cos(x^4)$

b. $p(x) = \sqrt{\tan(x)}$

c. $s(x) = 2^{\sin(x)}$

d. $z(x) = \cot^5(x)$

e. $m(x) = (\sec(x) + e^x)^9$

Activity 2.5.3. For each of the following functions, find the function's derivative. State the rule(s) you use, label relevant derivatives appropriately, and be sure to clearly identify your overall answer.

a. $p(r) = 4\sqrt{r^6 + 2e^r}$

d. $s(z) = 2^{z^2 \sec(z)}$

b. $m(v) = \sin(v^2)\cos(v^3)$

c. $h(y) = \dfrac{\cos(10y)}{e^{4y}+1}$

e. $c(x) = \sin(e^{x^2})$

Activity 2.5.4. Use known derivative rules, including the chain rule, as needed to answer each of the following questions.

    a. Find an equation for the tangent line to the curve $y = \sqrt{e^x + 3}$ at the point where $x = 0$.

    b. If $s(t) = \dfrac{1}{(t^2 + 1)^3}$ represents the position function of a particle moving horizontally along an axis at time t (where s is measured in inches and t in seconds), find the particle's instantaneous velocity at $t = 1$. Is the particle moving to the left or right at that instant?

    c. At sea level, air pressure is 30 inches of mercury. At an altitude of h feet above sea level, the air pressure, P, in inches of mercury, is given by the function $P = 30e^{-0.0000323h}$. Compute dP/dh and explain what this derivative function tells you about air pressure, including a discussion of the units on dP/dh. In addition, determine how fast the air pressure is changing for a pilot of a small plane passing through an altitude of 1000 feet.

    d. Suppose that $f(x)$ and $g(x)$ are differentiable functions and that the following information about them is known:

x	$f(x)$	$f'(x)$	$g(x)$	$g'(x)$
-1	2	-5	-3	4
2	-3	4	-1	2

Table 2.5.6: Data for functions f and g.

If $C(x)$ is a function given by the formula $f(g(x))$, determine $C'(2)$. In addition, if $D(x)$ is the function $f(f(x))$, find $D'(-1)$.

2.6 Derivatives of Inverse Functions

Preview Activity 2.6.1. The equation $y = \frac{5}{9}(x - 32)$ relates a temperature given in x degrees Fahrenheit to the corresponding temperature y measured in degrees Celcius.

    a. Solve the equation $y = \frac{5}{9}(x - 32)$ for x to write x (Fahrenheit temperature) in terms of y (Celcius temperature).

    b. Let $C(x) = \frac{5}{9}(x - 32)$ be the function that takes a Fahrenheit temperature as input and produces the Celcius temperature as output. In addition, let $F(y)$ be the function that converts a temperature given in y degrees Celcius to the temperature $F(y)$ measured in degrees Fahrenheit. Use your work in (a) to write a formula for $F(y)$.

    c. Next consider the new function defined by $p(x) = F(C(x))$. Use the formulas for F and C to determine an expression for $p(x)$ and simplify this expression as much as possible. What do you observe?

    d. Now, let $r(y) = C(F(y))$. Use the formulas for F and C to determine an expression for $r(y)$ and simplify this expression as much as possible. What do you observe?

    e. What is the value of $C'(x)$? of $F'(y)$? How do these values appear to be related?

Activity 2.6.2. For each function given below, find its derivative.

a. $h(x) = x^2 \ln(x)$

b. $p(t) = \frac{\ln(t)}{e^t + 1}$

c. $s(y) = \ln(\cos(y) + 2)$

d. $z(x) = \tan(\ln(x))$

e. $m(z) = \ln(\ln(z))$

Activity 2.6.3. The following prompts in this activity will lead you to develop the derivative of the inverse tangent function.

    a. Let $r(x) = \arctan(x)$. Use the relationship between the arctangent and tangent functions to rewrite this equation using only the tangent function.

    b. Differentiate both sides of the equation you found in (a). Solve the resulting equation for $r'(x)$, writing $r'(x)$ as simply as possible in terms of a trigonometric function evaluated at $r(x)$.

    c. Recall that $r(x) = \arctan(x)$. Update your expression for $r'(x)$ so that it only involves trigonometric functions and the independent variable x.

    d. Introduce a right triangle with angle θ so that $\theta = \arctan(x)$. What are the three sides of the triangle?

    e. In terms of only x and 1, what is the value of $\cos(\arctan(x))$?

    f. Use the results of your work above to find an expression involving only 1 and x for $r'(x)$.

Activity 2.6.4. Determine the derivative of each of the following functions.

a. $f(x) = x^3 \arctan(x) + e^x \ln(x)$

b. $p(t) = 2^{t \arcsin(t)}$

c. $h(z) = (\arcsin(5z) + \arctan(4 - z))^{27}$

d. $s(y) = \cot(\arctan(y))$

e. $m(v) = \ln(\sin^2(v) + 1)$

f. $g(w) = \arctan\left(\dfrac{\ln(w)}{1 + w^2}\right)$

2.7 Derivatives of Functions Given Implicitly

Preview Activity 2.7.1. Let f be a differentiable function of x (whose formula is not known) and recall that $\frac{d}{dx}[f(x)]$ and $f'(x)$ are interchangeable notations. Determine each of the following derivatives of combinations of explicit functions of x, the unknown function f, and an arbitrary constant c.

a. $\frac{d}{dx}\left[x^2 + f(x)\right]$

b. $\frac{d}{dx}\left[x^2 f(x)\right]$

c. $\frac{d}{dx}\left[c + x + f(x)^2\right]$

d. $\frac{d}{dx}\left[f(x^2)\right]$

e. $\frac{d}{dx}\left[xf(x) + f(cx) + cf(x)\right]$

Activity 2.7.2. Consider the curve defined by the equation $x = y^5 - 5y^3 + 4y$, whose graph is pictured in Figure 2.7.5.

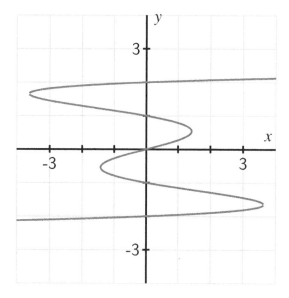

a. Explain why it is not possible to express y as an explicit function of x.

b. Use implicit differentiation to find a formula for dy/dx.

c. Use your result from part (b) to find an equation of the line tangent to the graph of $x = y^5 - 5y^3 + 4y$ at the point $(0, 1)$.

d. Use your result from part (b) to determine all of the points at which the graph of $x = y^5 - 5y^3 + 4y$ has a vertical tangent line.

Figure 2.7.5: The curve $x = y^5 - 5y^3 + 4y$.

Activity 2.7.3. Consider the curve defined by the equation $y(y^2 - 1)(y - 2) = x(x - 1)(x - 2)$, whose graph is pictured in Figure 2.7.6. Through implicit differentiation, it can be shown that

$$\frac{dy}{dx} = \frac{(x - 1)(x - 2) + x(x - 2) + x(x - 1)}{(y^2 - 1)(y - 2) + 2y^2(y - 2) + y(y^2 - 1)}.$$

Use this fact to answer each of the following questions.

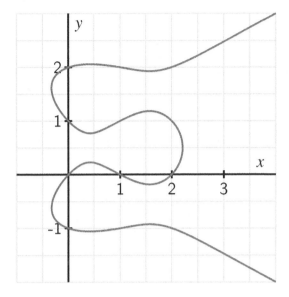

a. Determine all points (x, y) at which the tangent line to the curve is horizontal. (Use technology appropriately to find the needed zeros of the relevant polynomial function.)

b. Determine all points (x, y) at which the tangent line is vertical. (Use technology appropriately to find the needed zeros of the relevant polynomial function.)

c. Find the equation of the tangent line to the curve at one of the points where $x = 1$.

Figure 2.7.6: $y(y^2 - 1)(y - 2) = x(x - 1)(x - 2)$.

Activity 2.7.4. For each of the following curves, use implicit differentiation to find dy/dx and determine the equation of the tangent line at the given point.

a. $x^3 - y^3 = 6xy$, $(-3, 3)$

c. $3xe^{-xy} = y^2$, $(0.619061, 1)$

b. $\sin(y) + y = x^3 + x$, $(0, 0)$

2.8 Using Derivatives to Evaluate Limits

Preview Activity 2.8.1. Let h be the function given by $h(x) = \frac{x^5+x-2}{x^2-1}$.

a. What is the domain of h?

b. Explain why $\displaystyle\lim_{x\to 1} \frac{x^5 + x - 2}{x^2 - 1}$ results in an indeterminate form.

c. Next we will investigate the behavior of both the numerator and denominator of h near the point where $x = 1$. Let $f(x) = x^5 + x - 2$ and $g(x) = x^2 - 1$. Find the local linearizations of f and g at $a = 1$, and call these functions $L_f(x)$ and $L_g(x)$, respectively.

d. Explain why $h(x) \approx \frac{L_f(x)}{L_g(x)}$ for x near $a = 1$.

e. Using your work from (c) and (d), evaluate

$$\lim_{x\to 1} \frac{L_f(x)}{L_g(x)}.$$

What do you think your result tells us about $\lim_{x\to 1} h(x)$?

f. Investigate the function $h(x)$ graphically and numerically near $x = 1$. What do you think is the value of $\lim_{x\to 1} h(x)$?

Activity 2.8.2. Evaluate each of the following limits. If you use L'Hôpital's Rule, indicate where it was used, and be certain its hypotheses are met before you apply it.

a. $\lim_{x \to 0} \frac{\ln(1+x)}{x}$

c. $\lim_{x \to 1} \frac{2\ln(x)}{1-e^{x-1}}$

b. $\lim_{x \to \pi} \frac{\cos(x)}{x}$

d. $\lim_{x \to 0} \frac{\sin(x)-x}{\cos(2x)-1}$

Activity 2.8.3. In this activity, we reason graphically from the following figure to evaluate limits of ratios of functions about which some information is known.

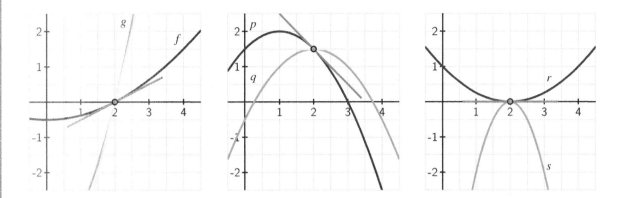

Figure 2.8.3: Three graphs referenced in the questions of Activity 2.8.3.

a. Use the left-hand graph to determine the values of $f(2)$, $f'(2)$, $g(2)$, and $g'(2)$. Then, evaluate $\lim\limits_{x \to 2} \frac{f(x)}{g(x)}$.

b. Use the middle graph to find $p(2)$, $p'(2)$, $q(2)$, and $q'(2)$. Then, determine the value of $\lim\limits_{x \to 2} \frac{p(x)}{q(x)}$.

c. Assume that r and s are functions whose for which $r''(2) \neq 0$ and $s''(2) \neq 0$ Use the right-hand graph to compute $r(2)$, $r'(2)$, $s(2)$, $s'(2)$. Explain why you cannot determine the exact value of $\lim\limits_{x \to 2} \frac{r(x)}{s(x)}$ without further information being provided, but that you can determine the sign of $\lim\limits_{x \to 2} \frac{r(x)}{s(x)}$. In addition, state what the sign of the limit will be, with justification.

Activity 2.8.4. Evaluate each of the following limits. If you use L'Hôpital's Rule, indicate where it was used, and be certain its hypotheses are met before you apply it.

a. $\lim_{x\to\infty} \frac{x}{\ln(x)}$

b. $\lim_{x\to\infty} \frac{e^x+x}{2e^x+x^2}$

c. $\lim_{x\to 0^+} \frac{\ln(x)}{\frac{1}{x}}$

d. $\lim_{x\to\frac{\pi}{2}^-} \frac{\tan(x)}{x-\frac{\pi}{2}}$

e. $\lim_{x\to\infty} xe^{-x}$

CHAPTER 3

Using Derivatives

3.1 Using derivatives to identify extreme values

Preview Activity 3.1.1. Consider the function h given by the graph in Figure 3.1.4. Use the graph to answer each of the following questions.

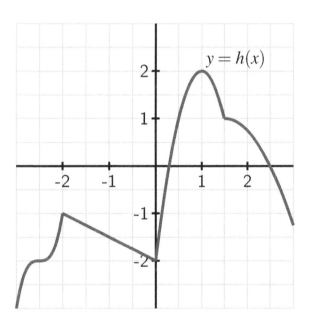

Figure 3.1.4: The graph of a function h on the interval $[-3, 3]$.

a. Identify all of the values of c such that $-3 < c < 3$ for which $h(c)$ is a local maximum of h.

b. Identify all of the values of c such that $-3 < c < 3$ for which $h(c)$ is a local minimum of h.

c. Does h have a global maximum on the interval $[-3, 3]$? If so, what is the value of this global maximum?

d. Does h have a global minimum on the interval $[-3, 3]$? If so, what is its value?

e. Identify all values of c for which $h'(c) = 0$.

f. Identify all values of c for which $h'(c)$ does not exist.

g. True or false: every relative maximum and minimum of h occurs at a point where $h'(c)$ is either zero or does not exist.

h. True or false: at every point where $h'(c)$ is zero or does not exist, h has a relative maximum or minimum.

Activity 3.1.2. Suppose that $g(x)$ is a function continuous for every value of $x \neq 2$ whose first derivative is $g'(x) = \frac{(x+4)(x-1)^2}{x-2}$. Further, assume that it is known that g has a vertical asymptote at $x = 2$.

   a. Determine all critical numbers of g.

   b. By developing a carefully labeled first derivative sign chart, decide whether g has as a local maximum, local minimum, or neither at each critical number.

   c. Does g have a global maximum? global minimum? Justify your claims.

   d. What is the value of $\lim_{x \to \infty} g'(x)$? What does the value of this limit tell you about the long-term behavior of g?

   e. Sketch a possible graph of $y = g(x)$.

Activity 3.1.3. Suppose that g is a function whose second derivative, g'', is given by the graph in Figure 3.1.15.

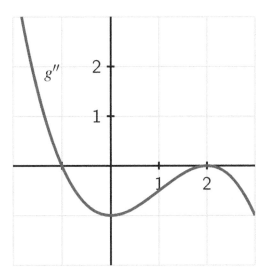

Figure 3.1.15: The graph of $y = g''(x)$.

a. Find the x-coordinates of all points of inflection of g.

b. Fully describe the concavity of g by making an appropriate sign chart.

c. Suppose you are given that $g'(-1.67857351) = 0$. Is there is a local maximum, local minimum, or neither (for the function g) at this critical number of g, or is it impossible to say? Why?

d. Assuming that $g''(x)$ is a polynomial (and that all important behavior of g'' is seen in the graph above), what degree polynomial do you think $g(x)$ is? Why?

Activity 3.1.4. Consider the family of functions given by $h(x) = x^2 + \cos(kx)$, where k is an arbitrary positive real number.

    a. Use a graphing utility to sketch the graph of h for several different k-values, including $k = 1, 3, 5, 10$. Plot $h(x) = x^2 + \cos(3x)$ on the axes provided. What is the smallest value of k at which you think you can see (just by looking at the graph) at least one inflection point on the graph of h?

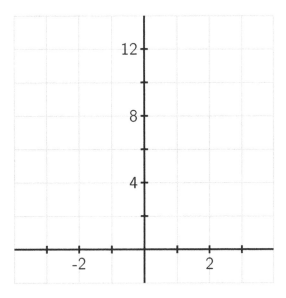

Figure 3.1.16: Axes for plotting $y = h(x)$.

    b. Explain why the graph of h has no inflection points if $k \leq \sqrt{2}$, but infinitely many inflection points if $k > \sqrt{2}$.

    c. Explain why, no matter the value of k, h can only have finitely many critical numbers.

3.2 Using derivatives to describe families of functions

Preview Activity 3.2.1. Let a, h, and k be arbitrary real numbers with $a \neq 0$, and let f be the function given by the rule $f(x) = a(x - h)^2 + k$.

    a. What familiar type of function is f? What information do you know about f just by looking at its form? (Think about the roles of a, h, and k.)

    b. Next we use some calculus to develop familiar ideas from a different perspective. To start, treat a, h, and k as constants and compute $f'(x)$.

    c. Find all critical numbers of f. (These will depend on at least one of a, h, and k.)

    d. Assume that $a < 0$. Construct a first derivative sign chart for f.

    e. Based on the information you've found above, classify the critical values of f as maxima or minima.

3.2 Using derivatives to describe families of functions

Activity 3.2.2. Consider the family of functions defined by $p(x) = x^3 - ax$, where $a \neq 0$ is an arbitrary constant.

    a. Find $p'(x)$ and determine the critical numbers of p. How many critical numbers does p have?

    b. Construct a first derivative sign chart for p. What can you say about the overall behavior of p if the constant a is positive? Why? What if the constant a is negative? In each case, describe the relative extremes of p.

    c. Find $p''(x)$ and construct a second derivative sign chart for p. What does this tell you about the concavity of p? What role does a play in determining the concavity of p?

    d. Without using a graphing utility, sketch and label typical graphs of $p(x)$ for the cases where $a > 0$ and $a < 0$. Label all inflection points and local extrema.

    e. Finally, use a graphing utility to test your observations above by entering and plotting the function $p(x) = x^3 - ax$ for at least four different values of a. Write several sentences to describe your overall conclusions about how the behavior of p depends on a.

Activity 3.2.3. Consider the two-parameter family of functions of the form $h(x) = a(1 - e^{-bx})$, where a and b are positive real numbers.

  a. Find the first derivative and the critical numbers of h. Use these to construct a first derivative sign chart and determine for which values of x the function h is increasing and decreasing.

  b. Find the second derivative and build a second derivative sign chart. For which values of x is a function in this family concave up? concave down?

  c. What is the value of $\lim_{x \to \infty} a(1 - e^{-bx})$? $\lim_{x \to -\infty} a(1 - e^{-bx})$?

  d. How does changing the value of b affect the shape of the curve?

  e. Without using a graphing utility, sketch the graph of a typical member of this family. Write several sentences to describe the overall behavior of a typical function h and how this behavior depends on a and b.

Activity 3.2.4. Let $L(t) = \frac{A}{1+ce^{-kt}}$, where A, c, and k are all positive real numbers.

a. Observe that we can equivalently write $L(t) = A(1 + ce^{-kt})^{-1}$. Find $L'(t)$ and explain why L has no critical numbers. Is L always increasing or always decreasing? Why?

b. Given the fact that

$$L''(t) = Ack^2e^{-kt}\,\frac{ce^{-kt}-1}{(1+ce^{-kt})^3},$$

find all values of t such that $L''(t) = 0$ and hence construct a second derivative sign chart. For which values of t is a function in this family concave up? concave down?

c. What is the value of $\lim_{t\to\infty} \frac{A}{1+ce^{-kt}}$? $\lim_{t\to-\infty} \frac{A}{1+ce^{-kt}}$?

d. Find the value of $L(x)$ at the inflection point found in (b).

e. Without using a graphing utility, sketch the graph of a typical member of this family. Write several sentences to describe the overall behavior of a typical function L and how this behavior depends on A, c, and k number.

f. Explain why it is reasonable to think that the function $L(t)$ models the growth of a population over time in a setting where the largest possible population the surrounding environment can support is A.

3.3 Global Optimization

Preview Activity 3.3.1. Let $f(x) = 2 + \frac{3}{1+(x+1)^2}$.

a. Determine all of the critical numbers of f.

b. Construct a first derivative sign chart for f and thus determine all intervals on which f is increasing or decreasing.

c. Does f have a global maximum? If so, why, and what is its value and where is the maximum attained? If not, explain why.

d. Determine $\lim_{x \to \infty} f(x)$ and $\lim_{x \to -\infty} f(x)$.

e. Explain why $f(x) > 2$ for every value of x.

f. Does f have a global minimum? If so, why, and what is its value and where is the minimum attained? If not, explain why.

.

Activity 3.3.2. Let $g(x) = \frac{1}{3}x^3 - 2x + 2$.

a. Find all critical numbers of g that lie in the interval $-2 \le x \le 3$.

b. Use a graphing utility to construct the graph of g on the interval $-2 \le x \le 3$.

c. From the graph, determine the x-values at which the absolute minimum and absolute maximum of g occur on the interval $[-2, 3]$.

d. How do your answers change if we instead consider the interval $-2 \le x \le 2$?

e. What if we instead consider the interval $-2 \le x \le 1$?

Activity 3.3.3. Find the *exact* absolute maximum and minimum of each function on the stated interval.

a. $h(x) = xe^{-x}$, $[0, 3]$

b. $p(t) = \sin(t) + \cos(t)$, $[-\frac{\pi}{2}, \frac{\pi}{2}]$

c. $q(x) = \frac{x^2}{x-2}$, $[3, 7]$

d. $f(x) = 4 - e^{-(x-2)^2}$, $(-\infty, \infty)$

e. $h(x) = xe^{-ax}$, $[0, \frac{2}{a}]$ $(a > 0)$

f. $f(x) = b - e^{-(x-a)^2}$, $(-\infty, \infty)$, $a, b > 0$

Activity 3.3.4. A piece of cardboard that is 10×15 (each measured in inches) is being made into a box without a top. To do so, squares are cut from each corner of the box and the remaining sides are folded up. If the box needs to be at least 1 inch deep and no more than 3 inches deep, what is the maximum possible volume of the box? what is the minimum volume? Justify your answers using calculus.

a. Draw a labeled diagram that shows the given information. What variable should we introduce to represent the choice we make in creating the box? Label the diagram appropriately with the variable, and write a sentence to state what the variable represents.

b. Determine a formula for the function V (that depends on the variable in (a)) that tells us the volume of the box.

c. What is the domain of the function V? That is, what values of x make sense for input? Are there additional restrictions provided in the problem?

d. Determine all critical numbers of the function V.

e. Evaluate V at each of the endpoints of the domain and at any critical numbers that lie in the domain.

f. What is the maximum possible volume of the box? the minimum?

3.4 Applied Optimization

Preview Activity 3.4.1. According to U.S. postal regulations, the girth plus the length of a parcel sent by mail may not exceed 108 inches, where by "girth" we mean the perimeter of the smallest end. What is the largest possible volume of a rectangular parcel with a square end that can be sent by mail? What are the dimensions of the package of largest volume?

   a. Let x represent the length of one side of the square end and y the length of the longer side. Label these quantities appropriately on the image shown in Figure 3.4.1.

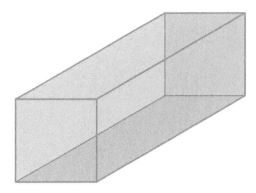

Figure 3.4.1: A rectangular parcel with a square end.

   b. What is the quantity to be optimized in this problem? Find a formula for this quantity in terms of x and y.

   c. The problem statement tells us that the parcel's girth plus length may not exceed 108 inches. In order to maximize volume, we assume that we will actually need the girth plus length to equal 108 inches. What equation does this produce involving x and y?

   d. Solve the equation you found in (c) for one of x or y (whichever is easier).

   e. Now use your work in (b) and (d) to determine a formula for the volume of the parcel so that this formula is a function of a single variable.

   f. Over what domain should we consider this function? Note that both x and y must be positive; how does the constraint that girth plus length is 108 inches produce intervals of possible values for x and y?

   g. Find the absolute maximum of the volume of the parcel on the domain you established in (f) and hence also determine the dimensions of the box of greatest volume. Justify that you've found the maximum using calculus.

Activity 3.4.2. A soup can in the shape of a right circular cylinder is to be made from two materials. The material for the side of the can costs $0.015 per square inch and the material for the lids costs $0.027 per square inch. Suppose that we desire to construct a can that has a volume of 16 cubic inches. What dimensions minimize the cost of the can?

    a. Draw a picture of the can and label its dimensions with appropriate variables.

    b. Use your variables to determine expressions for the volume, surface area, and cost of the can.

    c. Determine the total cost function as a function of a single variable. What is the domain on which you should consider this function?

    d. Find the absolute minimum cost and the dimensions that produce this value.

Activity 3.4.3. A hiker starting at a point P on a straight road walks east towards point Q, which is on the road and 3 kilometers from point P.

Two kilometers due north of point Q is a cabin. The hiker will walk down the road for a while, at a pace of 8 kilometers per hour. At some point Z between P and Q, the hiker leaves the road and makes a straight line towards the cabin through the woods, hiking at a pace of 3 kph, as pictured in Figure 3.4.3. In order to minimize the time to go from P to Z to the cabin, where should the hiker turn into the forest?

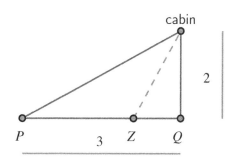

Figure 3.4.3: A hiker walks from P to Z to the cabin, as pictured.

Activity 3.4.4. Consider the region in the x-y plane that is bounded by the x-axis and the function $f(x) = 25 - x^2$. Construct a rectangle whose base lies on the x-axis and is centered at the origin, and whose sides extend vertically until they intersect the curve $y = 25 - x^2$. Which such rectangle has the maximum possible area? Which such rectangle has the greatest perimeter? Which has the greatest combined perimeter and area? (Challenge: answer the same questions in terms of positive parameters a and b for the function $f(x) = b - ax^2$.)

Activity 3.4.5. A trough is being constructed by bending a 4×24 (measured in feet) rectangular piece of sheet metal.

Two symmetric folds 2 feet apart will be made parallel to the longest side of the rectangle so that the trough has cross-sections in the shape of a trapezoid, as pictured in Figure 3.4.4. At what angle should the folds be made to produce the trough of maximum volume?

Figure 3.4.4: A cross-section of the trough formed by folding to an angle of θ.

3.5 Related Rates

Preview Activity 3.5.1. A spherical balloon is being inflated at a constant rate of 20 cubic inches per second. How fast is the radius of the balloon changing at the instant the balloon's diameter is 12 inches? Is the radius changing more rapidly when $d = 12$ or when $d = 16$? Why?

a. Draw several spheres with different radii, and observe that as volume changes, the radius, diameter, and surface area of the balloon also change.

b. Recall that the volume of a sphere of radius r is $V = \frac{4}{3}\pi r^3$. Note well that in the setting of this problem, *both* V and r are changing as time t changes, and thus both V and r may be viewed as implicit functions of t, with respective derivatives $\frac{dV}{dt}$ and $\frac{dr}{dt}$. Differentiate both sides of the equation $V = \frac{4}{3}\pi r^3$ with respect to t (using the chain rule on the right) to find a formula for $\frac{dV}{dt}$ that depends on both r and $\frac{dr}{dt}$.

c. At this point in the problem, by differentiating we have "related the rates" of change of V and r. Recall that we are given in the problem that the balloon is being inflated at a constant *rate* of 20 cubic inches per second. Is this rate the value of $\frac{dr}{dt}$ or $\frac{dV}{dt}$? Why?

d. From part (c), we know the value of $\frac{dV}{dt}$ at every value of t. Next, observe that when the diameter of the balloon is 12, we know the value of the radius. In the equation $\frac{dV}{dt} = 4\pi r^2 \frac{dr}{dt}$, substitute these values for the relevant quantities and solve for the remaining unknown quantity, which is $\frac{dr}{dt}$. How fast is the radius changing at the instant $d = 12$?

e. How is the situation different when $d = 16$? When is the radius changing more rapidly, when $d = 12$ or when $d = 16$?

Activity 3.5.2. A water tank has the shape of an inverted circular cone (point down) with a base of radius 6 feet and a depth of 8 feet. Suppose that water is being pumped into the tank at a constant instantaneous rate of 4 cubic feet per minute.

a. Draw a picture of the conical tank, including a sketch of the water level at a point in time when the tank is not yet full. Introduce variables that measure the radius of the water's surface and the water's depth in the tank, and label them on your figure.

b. Say that r is the radius and h the depth of the water at a given time, t. What equation relates the radius and height of the water, and why?

c. Determine an equation that relates the volume of water in the tank at time t to the depth h of the water at that time.

d. Through differentiation, find an equation that relates the instantaneous rate of change of water volume with respect to time to the instantaneous rate of change of water depth at time t.

e. Find the instantaneous rate at which the water level is rising when the water in the tank is 3 feet deep.

f. When is the water rising most rapidly: at $h = 3$, $h = 4$, or $h = 5$?

Activity 3.5.3. A television camera is positioned 4000 feet from the base of a rocket launching pad. The angle of elevation of the camera has to change at the correct rate in order to keep the rocket in sight. In addition, the auto-focus of the camera has to take into account the increasing distance between the camera and the rocket. We assume that the rocket rises vertically. (A similar problem is discussed and pictured dynamically at http://gvsu.edu/s/9t. Exploring the applet at the link will be helpful to you in answering the questions that follow.)

    a. Draw a figure that summarizes the given situation. What parts of the picture are changing? What parts are constant? Introduce appropriate variables to represent the quantities that are changing.

    b. Find an equation that relates the camera's angle of elevation to the height of the rocket, and then find an equation that relates the instantaneous rate of change of the camera's elevation angle to the instantaneous rate of change of the rocket's height (where all rates of change are with respect to time).

    c. Find an equation that relates the distance from the camera to the rocket to the rocket's height, as well as an equation that relates the instantaneous rate of change of distance from the camera to the rocket to the instantaneous rate of change of the rocket's height (where all rates of change are with respect to time).

    d. Suppose that the rocket's speed is 600 ft/sec at the instant it has risen 3000 feet. How fast is the distance from the television camera to the rocket changing at that moment? If the camera is following the rocket, how fast is the camera's angle of elevation changing at that same moment?

    e. If from an elevation of 3000 feet onward the rocket continues to rise at 600 feet/sec, will the rate of change of distance with respect to time be greater when the elevation is 4000 feet than it was at 3000 feet, or less? Why?

Activity 3.5.4. As pictured in the applet at http://gvsu.edu/s/9q, a skateboarder who is 6 feet tall rides under a 15 foot tall lamppost at a constant rate of 3 feet per second. We are interested in understanding how fast his shadow is changing at various points in time.

a. Draw an appropriate right triangle that represents a snapshot in time of the skateboarder, lamppost, and his shadow. Let x denote the horizontal distance from the base of the lamppost to the skateboarder and s represent the length of his shadow. Label these quantities, as well as the skateboarder's height and the lamppost's height on the diagram.

b. Observe that the skateboarder and the lamppost represent parallel line segments in the diagram, and thus similar triangles are present. Use similar triangles to establish an equation that relates x and s.

c. Use your work in (b) to find an equation that relates $\frac{dx}{dt}$ and $\frac{ds}{dt}$.

d. At what rate is the length of the skateboarder's shadow increasing at the instant the skateboarder is 8 feet from the lamppost?

e. As the skateboarder's distance from the lamppost increases, is his shadow's length increasing at an increasing rate, increasing at a decreasing rate, or increasing at a constant rate?

f. Which is moving more rapidly: the skateboarder or the tip of his shadow? Explain, and justify your answer.

Activity 3.5.5. A baseball diamond is 90′ square. A batter hits a ball along the third base line and runs to first base. At what rate is the distance between the ball and first base changing when the ball is halfway to third base, if at that instant the ball is traveling 100 feet/sec? At what rate is the distance between the ball and the runner changing at the same instant, if at the same instant the runner is 1/8 of the way to first base running at 30 feet/sec?

CHAPTER 4

The Definite Integral

4.1 Determining distance traveled from velocity

Preview Activity 4.1.1. Suppose that a person is taking a walk along a long straight path and walks at a constant rate of 3 miles per hour.

    a. On the left-hand axes provided in Figure 4.1.1, sketch a labeled graph of the velocity function $v(t) = 3$.

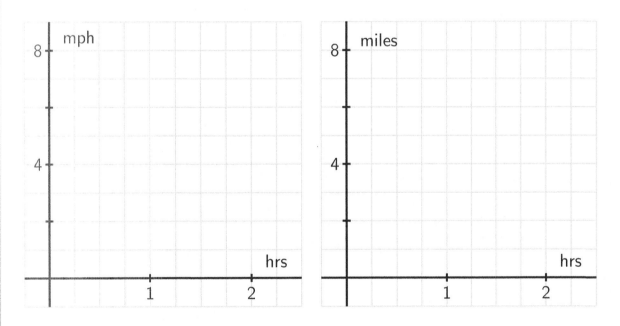

Figure 4.1.1: At left, axes for plotting $y = v(t)$; at right, for plotting $y = s(t)$.

    Note that while the scale on the two sets of axes is the same, the units on the right-hand axes differ from those on the left. The right-hand axes will be used in question (d).

    b. How far did the person travel during the two hours? How is this distance related to the area of a certain region under the graph of $y = v(t)$?

    c. Find an algebraic formula, $s(t)$, for the position of the person at time t, assuming that $s(0) = 0$. Explain your thinking.

    d. On the right-hand axes provided in Figure 4.1.1, sketch a labeled graph of the position function $y = s(t)$.

e. For what values of t is the position function s increasing? Explain why this is the case using relevant information about the velocity function v.

Activity 4.1.2. Suppose that a person is walking in such a way that her velocity varies slightly according to the information given in Table 4.1.4 and graph given in Figure 4.1.5.

t	$v(t)$
0.00	1.500
0.25	1.789
0.50	1.938
0.75	1.992
1.00	2.000
1.25	2.008
1.50	2.063
1.75	2.211
2.00	2.500

Table 4.1.4: Velocity data for the person walking.

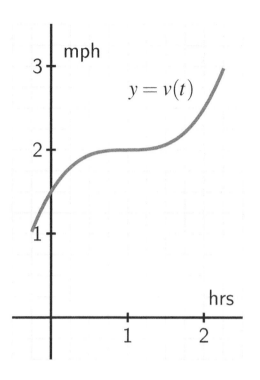

Figure 4.1.5: The graph of $y = v(t)$.

a. Using the grid, graph, and given data appropriately, estimate the distance traveled by the walker during the two hour interval from $t = 0$ to $t = 2$. You should use time intervals of width $\Delta t = 0.5$, choosing a way to use the function consistently to determine the height of each rectangle in order to approximate distance traveled.

b. How could you get a better approximation of the distance traveled on $[0, 2]$? Explain, and then find this new estimate.

c. Now suppose that you know that v is given by $v(t) = 0.5t^3 - 1.5t^2 + 1.5t + 1.5$. Remember that v is the derivative of the walker's position function, s. Find a formula for s so that $s' = v$.

d. Based on your work in (c), what is the value of $s(2) - s(0)$? What is the meaning of this quantity?

Activity 4.1.3. A ball is tossed vertically in such a way that its velocity function is given by $v(t) = 32 - 32t$, where t is measured in seconds and v in feet per second. Assume that this function is valid for $0 \leq t \leq 2$.

    a. For what values of t is the velocity of the ball positive? What does this tell you about the motion of the ball on this interval of time values?

    b. Find an antiderivative, s, of v that satisfies $s(0) = 0$.

    c. Compute the value of $s(1) - s(\frac{1}{2})$. What is the meaning of the value you find?

    d. Using the graph of $y = v(t)$ provided in Figure 4.1.8, find the exact area of the region under the velocity curve between $t = \frac{1}{2}$ and $t = 1$. What is the meaning of the value you find?

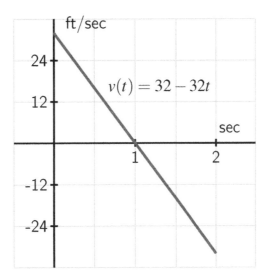

Figure 4.1.8: The graph of $y = v(t)$.

    e. Answer the same questions as in (c) and (d) but instead using the interval $[0, 1]$.

    f. What is the value of $s(2) - s(0)$? What does this result tell you about the flight of the ball? How is this value connected to the provided graph of $y = v(t)$? Explain.

Activity 4.1.4. Suppose that an object moving along a straight line path has its velocity v (in meters per second) at time t (in seconds) given by the piecewise linear function whose graph is pictured at left in Figure 4.1.10. We view movement to the right as being in the positive direction (with positive velocity), while movement to the left is in the negative direction.

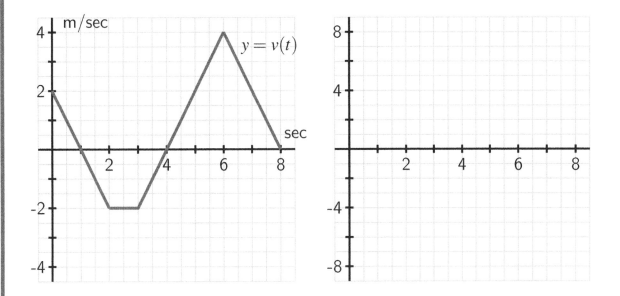

Figure 4.1.10: The velocity function of a moving object.

Suppose further that the object's initial position at time $t = 0$ is $s(0) = 1$.

a. Determine the total distance traveled and the total change in position on the time interval $0 \le t \le 2$. What is the object's position at $t = 2$?

b. On what time intervals is the moving object's position function increasing? Why? On what intervals is the object's position decreasing? Why?

c. What is the object's position at $t = 8$? How many total meters has it traveled to get to this point (including distance in both directions)? Is this different from the object's total change in position on $t = 0$ to $t = 8$?

d. Find the exact position of the object at $t = 1, 2, 3, \ldots, 8$ and use this data to sketch an accurate graph of $y = s(t)$ on the axes provided at right in Figure 4.1.10. How can you use the provided information about $y = v(t)$ to determine the concavity of s on each relevant interval?

4.2 Riemann Sums

Preview Activity 4.2.1. A person walking along a straight path has her velocity in miles per hour at time t given by the function $v(t) = 0.25t^3 - 1.5t^2 + 3t + 0.25$, for times in the interval $0 \leq t \leq 2$. The graph of this function is also given in each of the three diagrams in Figure 4.2.2.

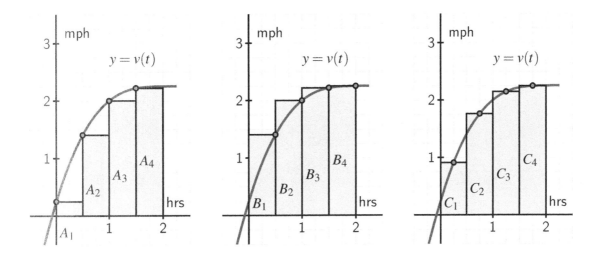

Figure 4.2.2: Three approaches to estimating the area under $y = v(t)$ on the interval $[0, 2]$.

Note that in each diagram, we use four rectangles to estimate the area under $y = v(t)$ on the interval $[0, 2]$, but the method by which the four rectangles' respective heights are decided varies among the three individual graphs.

a. How are the heights of rectangles in the left-most diagram being chosen? Explain, and hence determine the value of
$$S = A_1 + A_2 + A_3 + A_4$$
by evaluating the function $y = v(t)$ at appropriately chosen values and observing the width of each rectangle. Note, for example, that
$$A_3 = v(1) \cdot \frac{1}{2} = 2 \cdot \frac{1}{2} = 1.$$

b. Explain how the heights of rectangles are being chosen in the middle diagram and find the value of
$$T = B_1 + B_2 + B_3 + B_4.$$

c. Likewise, determine the pattern of how heights of rectangles are chosen in the right-most diagram and determine
$$U = C_1 + C_2 + C_3 + C_4.$$

d. Of the estimates S, T, and U, which do you think is the best approximation of D, the total distance the person traveled on $[0, 2]$? Why?

Activity 4.2.2. For each sum written in sigma notation, write the sum long-hand and evaluate the sum to find its value. For each sum written in expanded form, write the sum in sigma notation.

a. $\sum_{k=1}^{5}(k^2 + 2)$

d. $4 + 8 + 16 + 32 + \cdots + 256$

b. $\sum_{i=3}^{6}(2i - 1)$

c. $3 + 7 + 11 + 15 + \cdots + 27$

e. $\sum_{i=1}^{6} \frac{1}{2^i}$

Activity 4.2.3. Suppose that an object moving along a straight line path has its velocity in feet per second at time t in seconds given by $v(t) = \frac{2}{9}(t-3)^2 + 2$.

    a. Carefully sketch the region whose exact area will tell you the value of the distance the object traveled on the time interval $2 \le t \le 5$.

    b. Estimate the distance traveled on $[2, 5]$ by computing L_4, R_4, and M_4.

    c. Does averaging L_4 and R_4 result in the same value as M_4? If not, what do you think the average of L_4 and R_4 measures?

    d. For this question, think about an arbitrary function f, rather than the particular function v given above. If f is positive and increasing on $[a, b]$, will L_n over-estimate or under-estimate the exact area under f on $[a, b]$? Will R_n over- or under-estimate the exact area under f on $[a, b]$? Explain.

Activity 4.2.4. Suppose that an object moving along a straight line path has its velocity v (in feet per second) at time t (in seconds) given by

$$v(t) = \frac{1}{2}t^2 - 3t + \frac{7}{2}.$$

a. Compute M_5, the middle Riemann sum, for v on the time interval $[1, 5]$. Be sure to clearly identify the value of Δt as well as the locations of $t_0, t_1, \cdots, t_5$. In addition, provide a careful sketch of the function and the corresponding rectangles that are being used in the sum.

b. Building on your work in (a), estimate the total change in position of the object on the interval $[1, 5]$.

c. Building on your work in (a) and (b), estimate the total distance traveled by the object on $[1, 5]$.

d. Use appropriate computing technology[1] to compute M_{10} and M_{20}. What exact value do you think the middle sum eventually approaches as n increases without bound? What does that number represent in the physical context of the overall problem?

---

[2]For instance, consider the applet at http://gvsu.edu/s/a9 and change the function and adjust the locations of the blue points that represent the interval endpoints a and b.

4.3 The Definite Integral

Preview Activity 4.3.1. Consider the applet found at http://gvsu.edu/s/a9[1]. There, you will initially see the situation shown in Figure 4.3.2.

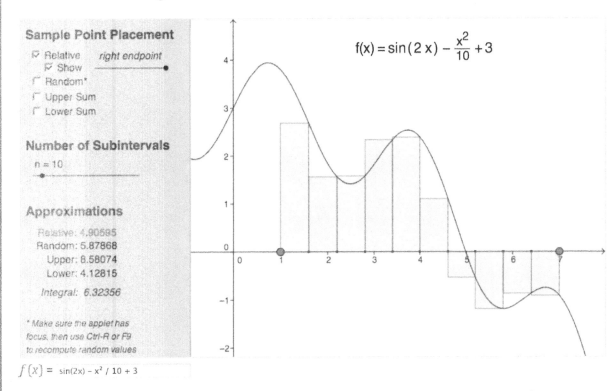

Sample Point Placement

☑ Relative   *right endpoint*
☑ Show   ————————————●
☐ Random*
☐ Upper Sum
☐ Lower Sum

Number of Subintervals

n = 10

●————————————

Approximations

Relative: 4.90595
Random: 5.87868
Upper: 8.58074
Lower: 4.12815

Integral: 6.32356

* Make sure the applet has focus, then use Ctrl-R or F9 to recompute random values

$f(x) = \sin(2x) - x^2 / 10 + 3$

$$f(x) = \sin(2x) - \frac{x^2}{10} + 3$$

Figure 4.3.2: A right Riemann sum with 10 subintervals for the function $f(x) = \sin(2x) - \frac{x^2}{10} + 3$ on the interval $[1, 7]$. The value of the sum is $R_{10} = 4.90595$.

Note that the value of the chosen Riemann sum is displayed next to the word "relative," and that you can change the type of Riemann sum being computed by dragging the point on the slider bar below the phrase "sample point placement."

Explore to see how you can change the window in which the function is viewed, as well as the function itself. You can set the minimum and maximum values of x by clicking and dragging on the blue points that set the endpoints; you can change the function by typing a new formula in the "f(x)" window at the bottom; and you can adjust the overall window by "panning and zooming" by using the Shift key and the scrolling feature of your mouse. More information on how to pan and zoom can be found at http://gvsu.edu/s/Fl.

Work accordingly to adjust the applet so that it uses a left Riemann sum with $n = 5$ subintervals for the function is $f(x) = 2x + 1$. You should see the updated figure shown in Figure 4.3.3. Then, answer the following questions.

a. Update the applet (and view window, as needed) so that the function being considered is $f(x) = 2x + 1$ on $[1, 4]$, as directed above. For this function on this interval, compute L_n, M_n, R_n for $n = 5$, $n = 25$, and $n = 100$. What appears to be the exact area bounded by $f(x) = 2x + 1$ and the x-axis on $[1, 4]$?

b. Use basic geometry to determine the exact area bounded by $f(x) = 2x + 1$ and the x-axis on $[1, 4]$.

c. Based on your work in (a) and (b), what do you observe occurs when we increase the number of subintervals used in the Riemann sum?

   d. Update the applet to consider the function $f(x) = x^2 + 1$ on the interval $[1, 4]$ (note that you need to enter "x ^ 2 + 1" for the function formula). Use the applet to compute L_n, M_n, R_n for $n = 5$, $n = 25$, and $n = 100$. What do you conjecture is the exact area bounded by $f(x) = x^2 + 1$ and the x-axis on $[1, 4]$?

   e. Why can we not compute the exact value of the area bounded by $f(x) = x^2 + 1$ and the x-axis on $[1, 4]$ using a formula like we did in (b)?

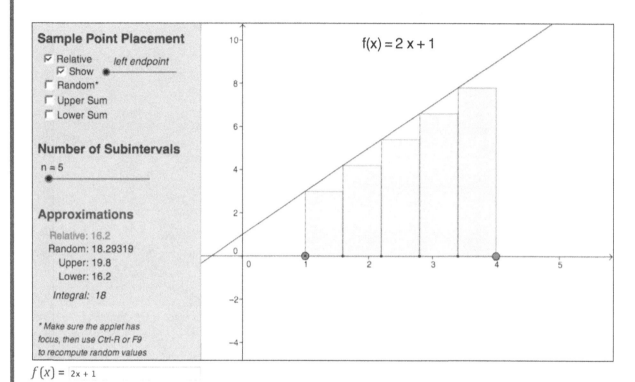

Figure 4.3.3: A left Riemann sum with 5 subintervals for the function $f(x) = 2x + 1$ on the interval $[1, 4]$. The value of the sum is $L_5 = 16.2$.

[1]Marc Renault, Shippensburg University, Geogebra Applets for Calclulus, http://gvsu.edu/s/5p.

Activity 4.3.2. Use known geometric formulas and the net signed area interpretation of the definite integral to evaluate each of the definite integrals below.

a. $\int_0^1 3x\,dx$

b. $\int_{-1}^4 (2 - 2x)\,dx$

c. $\int_{-1}^1 \sqrt{1 - x^2}\,dx$

d. $\int_{-3}^4 g(x)\,dx$, where g is the function pictured in Figure 4.3.7. Assume that each portion of g is either part of a line or part of a circle.

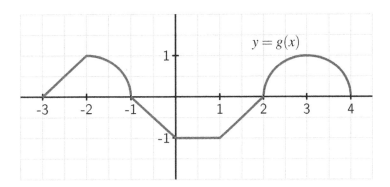

Figure 4.3.7: A function g that is piecewise defined; each piece of the function is part of a circle or part of a line.

Activity 4.3.3. Suppose that the following information is known about the functions f, g, x^2, and x^3:

- $\int_0^2 f(x)\,dx = -3$; $\int_2^5 f(x)\,dx = 2$

- $\int_0^2 g(x)\,dx = 4$; $\int_2^5 g(x)\,dx = -1$

- $\int_0^2 x^2\,dx = \frac{8}{3}$; $\int_2^5 x^2\,dx = \frac{117}{3}$

- $\int_0^2 x^3\,dx = 4$; $\int_2^5 x^3\,dx = \frac{609}{4}$

Use the provided information and the rules discussed in the preceding section to evaluate each of the following definite integrals.

a. $\int_5^2 f(x)\,dx$

b. $\int_0^5 g(x)\,dx$

c. $\int_0^5 (f(x) + g(x))\,dx$

d. $\int_2^5 (3x^2 - 4x^3)\,dx$

e. $\int_5^0 (2x^3 - 7g(x))\,dx$

191

Activity 4.3.4. Suppose that $v(t) = \sqrt{4 - (t-2)^2}$ tells us the instantaneous velocity of a moving object on the interval $0 \leq t \leq 4$, where t is measured in minutes and v is measured in meters per minute.

a. Sketch an accurate graph of $y = v(t)$. What kind of curve is $y = \sqrt{4 - (t-2)^2}$?

b. Evaluate $\int_0^4 v(t)\,dt$ exactly.

c. In terms of the physical problem of the moving object with velocity $v(t)$, what is the meaning of $\int_0^4 v(t)\,dt$? Include units on your answer.

d. Determine the exact average value of $v(t)$ on $[0,4]$. Include units on your answer.

e. Sketch a rectangle whose base is the line segment from $t = 0$ to $t = 4$ on the t-axis such that the rectangle's area is equal to the value of $\int_0^4 v(t)\,dt$. What is the rectangle's exact height?

f. How can you use the average value you found in (d) to compute the total distance traveled by the moving object over $[0,4]$?

4.4 The Fundamental Theorem of Calculus

Preview Activity 4.4.1. A student with a third floor dormitory window 32 feet off the ground tosses a water balloon straight up in the air with an initial velocity of 16 feet per second. It turns out that the instantaneous velocity of the water balloon is given by $v(t) = -32t + 16$, where v is measured in feet per second and t is measured in seconds.

a. Let $s(t)$ represent the height of the water balloon above ground at time t, and note that s is an antiderivative of v. That is, v is the derivative of s: $s'(t) = v(t)$. Find a formula for $s(t)$ that satisfies the initial condition that the balloon is tossed from 32 feet above ground. In other words, make your formula for s satisfy $s(0) = 32$.

b. When does the water balloon reach its maximum height? When does it land?

c. Compute $s(\frac{1}{2}) - s(0)$, $s(2) - s(\frac{1}{2})$, and $s(2) - s(0)$. What do these represent?

d. What is the total vertical distance traveled by the water balloon from the time it is tossed until the time it lands?

e. Sketch a graph of the velocity function $y = v(t)$ on the time interval $[0, 2]$. What is the total net signed area bounded by $y = v(t)$ and the t-axis on $[0, 2]$? Answer this question in two ways: first by using your work above, and then by using a familiar geometric formula to compute areas of certain relevant regions.

Activity 4.4.2. Use the Fundamental Theorem of Calculus to evaluate each of the following integrals exactly. For each, sketch a graph of the integrand on the relevant interval and write one sentence that explains the meaning of the value of the integral in terms of the (net signed) area bounded by the curve.

a. $\int_{-1}^{4}(2-2x)\,dx$

b. $\int_{0}^{\frac{\pi}{2}}\sin(x)\,dx$

c. $\int_{0}^{1}e^{x}\,dx$

d. $\int_{-1}^{1}x^{5}\,dx$

e. $\int_{0}^{2}(3x^{3}-2x^{2}-e^{x})\,dx$

Activity 4.4.3. Use your knowledge of derivatives of basic functions to complete Table 4.4.5 of antiderivatives. For each entry, your task is to find a function F whose derivative is the given function f. When finished, use the FTC and the results in the table to evaluate the three given definite integrals.

given function, $f(x)$	antiderivative, $F(x)$
k, (k is constant)	
x^n, $n \neq -1$	
$\frac{1}{x}$, $x > 0$	
$\sin(x)$	
$\cos(x)$	
$\sec(x)\tan(x)$	
$\csc(x)\cot(x)$	
$\sec^2(x)$	
$\csc^2(x)$	
e^x	
a^x $(a > 1)$	
$\frac{1}{1+x^2}$	
$\frac{1}{\sqrt{1-x^2}}$	

Table 4.4.5: Familiar basic functions and their antiderivatives.

a. $\displaystyle\int_0^1 \left(x^3 - x - e^x + 2\right)\, dx$

b. $\displaystyle\int_0^{\pi/3} \left(2\sin(t) - 4\cos(t) + \sec^2(t) - \pi\right) dt$

c. $\displaystyle\int_0^1 \left(\sqrt{x} - x^2\right) dx$

Activity 4.4.4. During a 40-minute workout, a person riding an exercise machine burns calories at a rate of c calories per minute, where the function $y = c(t)$ is given in Figure 4.4.9. On the interval $0 \le t \le 10$, the formula for c is $c(t) = -0.05t^2 + t + 10$, while on $30 \le t \le 40$, its formula is $c(t) = -0.05t^2 + 3t - 30$.

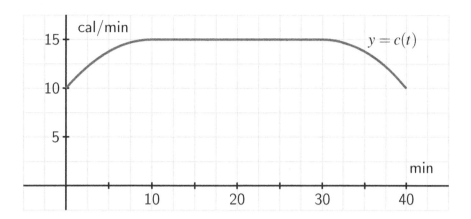

Figure 4.4.9: The rate $c(t)$ at which a person exercising burns calories, measured in calories per minute.

a. What is the exact total number of calories the person burns during the first 10 minutes of her workout?

b. Let $C(t)$ be an antiderivative of $c(t)$. What is the meaning of $C(40) - C(0)$ in the context of the person exercising? Include units on your answer.

c. Determine the exact average rate at which the person burned calories during the 40-minute workout.

d. At what time(s), if any, is the instantaneous rate at which the person is burning calories equal to the average rate at which she burns calories, on the time interval $0 \le t \le 40$?

Colophon

This book was authored in PreTeXt.

Made in the USA
Monee, IL
24 July 2023

39832784R00116